KETTERING COLLEGE
MEDICAL ARTS LIBRARY

D0145925

Wondering About Physics . . .
Using Spreadsheets to Find Out

Wondering About Physics . . .
Using Spreadsheets to Find Out

Investigations in Physics

Dewey I. Dykstra, Jr.
Boise State University

Robert G. Fuller
United States Air Force Academy
University of Nebraska–Lincoln

WILEY

John Wiley & Sons
New York • Chichester • Brisbane • Toronto • Singapore

Copyright © 1988 by John Wiley & Sons, Inc.

All rights reserved. Published simultaneously in Canada.

Reproduction or translation of any part of
this work beyond that permitted by Sections
107 or 108 of the 1976 United States Copyright
Act without the permission of the copyright
owner is unlawful. Requests for permission or
further information should be addressed to the
Permissions Department, John Wiley & Sons, Inc.

Library of Congress Cataloging in Publication Data:

Dykstra, Dewey I. (Dewey Irwin), 1947–
    Wondering about physics : using spreadsheets to find out
  investigations in physics / Dewey I. Dykstra, Jr., Robert G. Fuller.
      p.   cm.
    Bibliography: p.
    ISBN 0-471-63174-4 (pbk.)
    1. Physics--Data processing. 2. Electronic Spreadsheets.
  I. Fuller, Robert G. II. Title.
  QC52.D95 1988
  530'.028'5--dc19   88-14895
                                                        CIP

Printed in the United States of America

10 9 8 7 6 5 4 3 2 1

All the figures signed with *were drawn by Andrew S. Fuller.*

Preface

Welcome to what we believe will be a new experience in exploring physics. This book contains a series of investigations in physics that are intended to accompany a standard textbook in physics. The introductory physics texts published by John Wiley & Sons will do nicely.

You will notice that while these investigations deal with the same topics as problems in the textbooks, they do so in a much different way from the problems and exercises at the ends of the chapters. We use the term "investigation" to convey the exploratory nature of the activities. The goal of each investigation is not to find "THE ANSWER," but to create a tool to explore the behavior of some physical situation. The nature of these explorations is sufficiently different from normal physics problems that we found numerous surprises and interesting twists to the way things work. We hope that you find some of the investigations as interesting as we have.

The sequence of these investigations is in the order that the topics occur in the typical physics text. The investigations are not in order of easy-to-hard, nor is the level of hints uniform throughout. We have used the following code in the Table of Contents to identify some characteristics of the investigations.

B Beginner level spreadsheet activity

M Mathematical Methods section or other extra assistance included with advice on useful techniques

D Difficult problem solving

C Complex physics concepts

Although we have a computer system that we would recomend if you asked us, we realize that one does not always have choices in these matters. We have written the investigations so that almost any computer running almost any spreadsheet software will do. Your spreadsheet must be able to do trigonometry functions (sine, cosine, tangent, and their inverses) and exponential and logarithmic (base e) functions and it would be best if it were integrated with software to graph the results of your calculations. None of the investigations will involve extremely large memory requirements.

If you are taking a physics course, your instructor may be providing handouts that help you get started with the computer and software available at your school. If not, do not be discouraged, the software usually comes with tutorials to teach you how to use it, and the major software products usually have a number of books available that are intended to help you learn to use the software. Sometimes these books are better than the tutorials provided by the software companies.

We have two purposes in writing this series of investigations:

1. To provide you with a setting to explore physics in a way that is free from the limitations imposed by the need for calculus. Although it is good to know calculus and to be able to use it, we believe that it is equally important to use powerful alternatives (the numerical methods available on the computer) to explore physics. Some interesting physical situations cannot be represented in closed form anyway. Numerical methods must be used in these cases.

2. To introduce you to a powerful tool for exploring the world of physics and for use in your future career. To be competitive in the future, the computer with its software tools must become for you a natural alternative in your "professional toolbox." Keep in mind that in 5 or 10 years from when you read this, spreadsheets may have been replaced by some other software tool that is more powerful and sophisticated, but you will not be equipped to take advantage of that more powerful software if spreadsheets have not already become old friends first.

Many people have contributed in one way or another to the writing of this book. We acknowledge Robert Tinker, Leslie Eiser, Alex Burr, Bill Evenson, Fred Loxsom, Frank Griffin, Marvin DeJong, John Robson, Harvey Leff, George Hept, Gary Lorenzen, and Tom Gist as some of those who have influenced our thinking. Of course, for errors and shortcomings we take all the blame.

We hope that you enjoy these investigations and discover that while exploring physics is sometimes hard, the fun is as much in running the race as in crossing the finish line.

Dewey I. Dykstra, Jr.
Robert G. Fuller

Contents

Comment Codes:

B   Beginner Level Spreadsheet Activity

M   Mathematical Methods Section or Other Extra Assistance Included with Advice on Useful Techniques

D   Difficult Problem Solving

C   Complex Physics Concepts

Investigation 1

REACTION TIMES

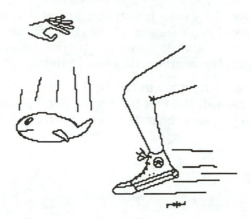

I. Wondering About...

...hand and foot reaction times.

II. Investigation to Undertake

Build a spreadsheet in which you can collect data about reaction times for various uses of your hands and feet: catching a dropped object with your foot, moving your hand away so that a small dropped object does not hit it, or moving your foot away so that an object does not hit it. These data should be collected on a number of people so that averages can be generated.

III. Background

Many factors affect your reaction time in using your hands and feet: the inertia of the limb being used, the speed of nerve signals from the brain to the involved muscles, the complexity of the action, the precision of the action, general muscle tone, and so on. Some of these reaction times are quite important in everyday life, say, in driving a car, for instance. The distance that you travel at various speeds while you are reacting to a situation can be greater than the distance you travel while you are applying the brakes or steering to avoid an unpleasant situation.

Reaction times are 0.10 sec or greater. How far does one travel at 60 mph in a 0.20 sec? In another setting, it is useful to realize that a golf ball is 9 m away by the time you can react to feeling it hit the club! Although typical impact times in golf are 0.0005 sec, in other sports impact times can be as long as 0.005 sec. You can use this and an estimate of the speed of the object being hit to estimate how far away it is by the time that you can react.

IV. Related Science Concepts

Use the time for an object to fall a short distance as your measure of reaction time. A meter stick or ruler will be required to measure the distance fallen.

1

1. For a simple hand reaction, place the subject's palm down on a table with the ruler standing straight up just beyond the finger tips. Drop a small object (a ball, paper clip, and so on) from increasing heights starting at about 5 cm (2 in.) until the subject can just get his or her fingers out of the way before the object hits the table. Does the thickness of the fingers matter?

2. For a more complex hand reaction, have the subject place his or her fore arm on the table with the hand extending out beyond the edge. The hand should be poised to grasp a vertically falling ruler. Hold the edge of the ruler just above the hand. Drop the ruler. How much of the ruler falls through the hand is a measure of the time for the ruler to fall while the hand is reacting. Just where on the hand do you measure the amount of fall of the ruler? Which reaction time is shorter on the average, this one or the previous one?

3. For a foot reaction, use an arrangement on the floor much like number 1 above. Place the ruler just beyond the toes with the foot flat on the floor. Drop a small object from various heights until the foot can just clear before the object can hit it. Does the thickness of the toes or shoe matter?

EQUATIONS

When falling from rest the distance fallen is

$$d = \frac{at^2}{2},$$

where d is the distance fallen, t is the time of fall, and a is the acceleration of the fall. The time to fall is therefore

$$t = \sqrt{\frac{2d}{a}}.$$

The acceleration of a falling object is 9.8 m/s² or 32 ft/s².

V. Mathematical Methods

Set up your spreadsheet with

1. The first column to list the subject's last name.

2. The second column to list the subject's first name.

3. The third column to indicate the distance fallen for the simple hand reaction.

4. The fourth column to calculate the time of fall for the simple hand reaction.

5. The fifth column to indicate the distance fallen for the complex hand reaction.

6. The sixth column to calculate the time of fall for the complex hand reaction.

7. The seventh column to indicate the distance fallen for the foot reaction.

8. The eighth column to calculate the time of fall for the foot reaction.

Put a label at the heading of each column and for measured and calculated values indicate the units used. At the bottom of each measured and calculated column have the spreadsheet calculate the average, maximum, and minimum values in the column.

Investigation 2

AIRPLANE FLIGHT TIMES

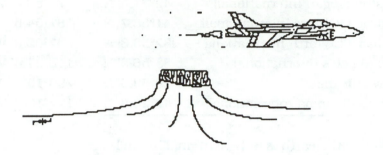

I. Wondering About...

...why it seems to take longer to fly from Atlanta to Los Angeles than it does to fly from Los Angeles to Atlanta.

II. Investigation to Undertake

Set up a spreadsheet to calculate average flight speeds. Why do they seem to differ depending on the direction of flight? Use your spreadsheet to develop a model to explain this phenomena.

III. Background

The scheduled times of flight as shown in the system timetable for a major commercial air line are as follows:

Flight Data

From	To	Miles	Flight Time (hr:min)
Atlanta	Seattle	2181	5:00
Seattle	Atlanta	2181	4:28
Chicago	New Orleans	837	2:01
New Orleans	Chicago	837	2:06
Atlanta	Los Angeles	1946	4:21
Los Angeles	Atlanta	1946	3:57
Atlanta	Dallas	731	2:00
Dallas	Atlanta	731	1:45
Atlanta	Boston	946	2:08
Boston	Atlanta	946	2:33

The locations of the airports as published in the U.S. Government Flight Information publication are as follows:

Airport Locations	Latitude	Longitude
Atlanta International	33°37.7' N	84°26.1' W
Boston (Logan International)	42°21.5' N	70°59.6' W
Chicago (O'Hare International)	41°58.7' N	87°54.5' W
Dallas–Ft. Worth Internationa	132°53.8' N	97°02.0' W
Los Angeles International	33°56.5' N	118°24.4' W
New Orleans	29°59.5' N	90°15.1' W
Seattle–Tacoma International	47°26.8' N	122°18.3' W

The mean radius of the earth is 6.37×10^6 m(3960 miles).

IV. Related Science Concepts

Velocity is a vector and the velocity of an airplane relative to the earth will depend upon both its velocity with respect to the air and the velocity of the air with respect to the ground. Use your knowledge of vectors to compute a velocity for air above the United States. Discuss the strengths and weaknesses of this model. How well does it fit the scheduled airlines data? How can you explain the discrepancies?

Investigation 3

OBJECTS FALLING FREELY FROM BUILDINGS

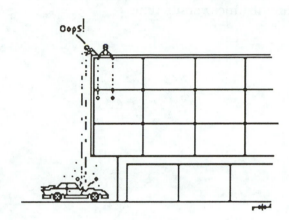

I. Wondering About...

...objects falling from buildings on the Earth, the Moon, Mars, Jupiter, and the Sun.

II. Investigation to Undertake

Set up a spreadsheet to compute the distance an object falls, as a function of the time of falling, near the surface of the Earth, the Moon, Mars, Jupiter, and the Sun. Compute the distance of fall each fifth of a second from 0 to 2 sec.

III. Background

All of us have dropped an object from the top of a tall building and watched it fall to the ground. How far does it fall in "one-thousand-and-one, one-thousand-and-two" seconds? How many stories tall is a building that requires an object to fall 2 sec from its top to reach the ground? How tall would it be on the Moon? Mars? Jupiter? The Sun? (1 story of a building approximately 3 meters.)

IV. Related Science Concepts

For a mass falling near the surface of a planet, its speed increases at a constant rate. This constant is called the acceleration caused by gravity, g. The numerical value of g is different for different objects in our solar system. The acceleration of a mass caused by the gravitational force at the surface for each of the following objects is: Earth 9.8 m/s^2; our Moon 1.6 m/s^2; Mars 3.8 m/s^2; Jupiter 25 m/s^2; and the Sun 270 m/s^2.

5

EQUATIONS

To compute the distance of fall, d:

$$d = \frac{gt^2}{2},$$

where g is the acceleration due to gravity and t is the time of fall. This equation can be used in the spreadsheet to compute distance versus time.

FALLING PARACHUTISTS

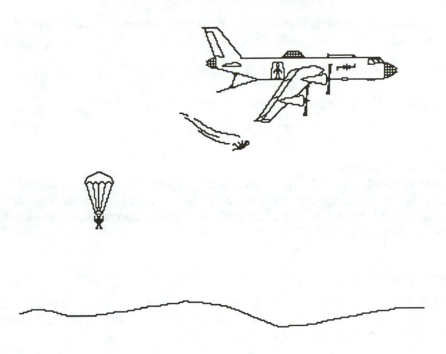

I. Wondering About...

...how fast a parachutist is falling after jumping from an airplane.

II. Investigation to Undertake

Set up a spreadsheet to compute how fast a 90-kg person is falling from an airplane as a function of the distance the person is below the altitude of the airplane. Consider three cases: falling with NO air resistance, falling with air resistance but NO parachute, and falling with a parachute open. Find the speed of the falling person every 20 m or so up to a distance of about 400 m.

III. Background

In treating idealized free falling objects near the surface of the earth, an object is assumed to fall with a constant acceleration of about 9.8 m/s². However, the presence of air does introduce a drag force that tends to reduce the actual acceleration of any real object falling in the air near the earth. This drag force is assumed to be opposite to the direction of the velocity of the object and have a magnitude proportional to the square of the magnitude of the velocity. For a typical human falling without a parachute the proportionality constant is about 0.23 N s²/m². Falling with a parachute increases the drag coefficient by about a factor of 100.

7

IV. Related Science Concepts

Consider the motion of a person in the vertical direction, falling straight down from rest. The net force on the person will be the sum of the force of gravitational attraction pulling down and the drag force acting against the fall, in an upward direction.

EQUATIONS

Newton's second law can be used: The sum of the external forces is equal to the mass of the object times its acceleration.

$$\sum F = ma ,$$

Take downward to be the positive direction, then

$$\sum F = mg - cv^2 = ma ,$$

where c is the drag coefficient.

This equation can be solved numerically with a spreadsheet using a process of iteration. Write a circular series of equations that refer to each other, and then have the spreadsheet calculate through the circle or the loop of equations until the numerical values in that row of the spreadsheet stop changing.

V. Mathematical Methods

Choose a constant increment of time Δt. Then the speed of the falling person at the end of the ith increment will be equal to the speed at the end of the $i^{th} - 1$ increment plus the acceleration during the ith time interval multiplied by Δt,

$$v_i = v_{i-1} + a_i \Delta t .$$

The acceleration during the ith time interval is equal to the difference between the ith value of speed minus the $i^{th} - 1$ value of the speed divided by Δt.

$$a_i = \frac{(v_i - v_{i-1})}{\Delta t}$$

For the special case of a person falling in the air with a drag force given by $-cv^2$, these equations for acceleration and velocity are related by

$$a_i = g - c \frac{\left[\frac{(v_i - v_{i-1})}{2} \right]^2}{m} .$$

where the term inside the square brackets is the average speed during the ith time interval.

The distance the person has fallen after the ith time interval can be computed using the standard definition of distance and speed,

$$x_i = x_{i-1} + \left[\frac{(v_i + v_{i-1})}{2} \right] \Delta t .$$

These equations can be used in a single row of a spreadsheet and, using the iteration process, solved numerically. (Note: This is an example of a simple problem that is difficult to solve exactly even using the power of differential equations.)

Investigation 5

THE LEANING TOWER OF PISA EXPERIMENT

I. Wondering about...

...what you would observe if you really dropped spheres from the top of the Leaning Tower of Pisa.

II. Investigation to Undertake

Set up a spreadsheet to compute the distance of fall from the top of the Leaning Tower of Pisa as a function of time for an oak sphere and a lead sphere, taking into account air resistance. Repeat the calculations for the Moon.

III. Background

The anecdotal history of physics reports that Galileo dropped two spheres from the top of the Leaning Tower of Pisa and they hit the ground at the same time, disproving the view of Aristotle. Thus, so the story goes, began the era of physics as we know it. Real history seems not to be in agreement with this apocryphal account of Galileo's experimental work. Nevertheless, pretend that you really did the real experiment both in Pisa and on the surface of the Moon. What would you observe?

9

IV. Related Science Concepts

An object falling freely near the surface of the earth will experience both the downward force of gravity and the drag force of the air resisting the motion of the object. Both of these forces need to be computed to find a realistic distance of fall for the spheres dropped from the Tower.

EQUATIONS

The net force which happens to be downward in this case:

$$\sum F = ma \,,$$

where m is the mass of the sphere and a is its acceleration, also downward.

The drag force, D, can be estimated to be proportional to the square of the velocity of the object falling through the air.

$$D = -cv^2,$$

where c is the drag coefficient and the minus sign indicates that this force is in the direction opposite to the direction of the velocity.

With the force due to gravity the net force can be expressed:

$$\sum F = mg - cv^2 = ma \,,$$

where, by implication, the downward direction is positive.

Take a typical value for the diameter of the spheres, say 10 cm. Consider the drag coefficient for air in Pisa to be about 2×10^{-3} N s^2/m^2. The top of the Tower of Pisa is about 47.8 m above the surface of the ground. Use 9.81 m/s^2 as the value for gravitational acceleration in Pisa and 1.67 m/s^2 for the Moon. The density of oak is about 0.74 gm/cm^3 and the density of lead is about 11.3 gm/cm^3.

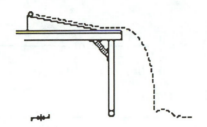

Investigation 6

UNDERSTANDING DATA
(MARBLES ROLLING OFF OF TABLES)

I. Wondering About...

...a study of the distances marbles roll off the edge of a table.

II. Investigation to Undertake

Build a short inclined ramp above a level table top. Let a marble roll from rest down the inclined ramp, across the table, and off the edge of the table onto the floor so that it lands on a piece of carbon paper placed on top of a sheet of graph paper printed with small squares. Repeat this about 40 times so that you obtain a distribution of carbon spots on the graph paper.

Transfer the carbon spot distribution to your spreadsheet by putting the number of spots in each square of the graph paper into a corresponding cell on the spreadsheet. Then use your spreadsheet to develop an understanding of various ways to analyze these data. Calculate averages and standard deviations for forward and sideways distances traveled from the table edge. Plot forward and sideways histograms. Compare your results to an ideal Gaussian distribution.

III. Background

When collecting experimental data, we discover that even when we repeat exactly, to the best of our ability, the same measurement over and over again, we do not get exactly the same numerical result. Inherent in an appropriate understanding of data is our ability to talk about the statistical nature of experimental data, using such terms as average value, most probable value, standard deviation, Gaussian distribution, histogram, distribution function, and so on.

IV. Related Science Concepts

Consider the situation where you have collected N independent data points, x_n, and you desire to make some inference about the "true" value of the quantity represented by x_n. One common approach in physics is to assume that the quantity x_n is drawn randomly from a single set of values with a mean of $\overline{x}$ and an uncertainty of $\pm\,\sigma$. You can then compute values of $\overline{x}$ and σ from the following equations.

11

EQUATIONS

$$\overline{x} = \frac{\sum x_n}{N} \text{ and } \sigma^2 = \frac{\sum (x_n - \overline{x})^2}{(N-1)},$$

where $\overline{x}$ is called the mean and σ is the standard deviation.

For a continuous probability distribution, the Gaussian distribution, also called the normal distribution, gives the probability that the value of x will lie between x and $x + \Delta x$ by the following equation:

$$P(x)\,\Delta x = \left[\frac{1}{(2\pi)^{\frac{1}{2}}\sigma} \right] \exp\left\{ \frac{-(x_n - \overline{x})^2}{2\sigma^2} \right\} \Delta x.$$

Some sample data for forty rolls of a marble:

Number of Marble Rolls	Velocity at Table Edge (m/s)	Angle of Launch (radians)	Time of Fall (sec)	Forward Distance (cm)	Sideways Distance (cm)
1	2.34	– 0.04	0.39	90.97	– 3.69
2	2.48	– 0.16	0.37	91.68	– 14.65
3	2.38	– 0.04	0.37	88.88	– 3.21
4	2.48	0.11	0.41	99.96	10.63
5	2.52	0.09	0.41	102.86	9.69
6	2.52	0.02	0.39	97.37	1.97
7	2.37	0.04	0.40	94.11	4.05
8	2.30	– 0.04	0.37	86.10	– 3.63
9	2.51	– 0.04	0.41	102.75	– 3.94
10	2.53	– 0.02	0.38	97.18	– 1.82
11	2.37	0.09	0.38	90.38	8.36
12	2.40	– 0.13	0.40	95.11	– 12.65
13	2.47	– 0.16	0.41	100.02	– 15.95
14	2.32	0.15	0.38	87.97	13.54
15	2.51	0.16	0.41	101.26	16.73
16	2.36	– 0.16	0.39	90.34	– 14.96
17	2.41	0.12	0.40	94.61	10.94
18	2.46	0.13	0.37	91.05	11.65
19	2.39	0.07	0.37	89.39	6.61
20	2.34	0.10	0.40	92.53	9.52
21	2.41	– 0.15	0.39	92.20	– 13.98
22	2.50	0.14	0.39	95.73	13.18
23	2.51	0.04	0.38	96.27	3.61
24	2.42	0.06	0.38	91.47	5.47
25	2.52	0.03	0.40	101.37	3.31
26	2.31	– 0.16	0.38	85.96	– 14.05
27	2.49	– 0.16	0.39	95.50	– 15.82
28	2.47	0.15	0.38	93.58	14.53
29	2.41	– 0.15	0.38	91.35	– 13.86
30	2.53	– 0.07	0.39	97.65	– 7.01
31	2.40	0.01	0.40	94.87	1.10
32	2.42	0.03	0.38	91.89	3.14
33	2.41	– 0.03	0.38	91.28	– 2.76
34	2.50	0.07	0.40	100.16	7.48
35	2.48	0.10	0.38	93.97	9.12
36	2.34	– 0.13	0.37	87.07	– 10.97
37	2.49	– 0.03	0.37	92.59	– 3.11
38	2.50	0.16	0.39	97.11	15.76
39	2.34	– 0.08	0.37	87.05	– 6.60
40	2.38	0.09	0.38	90.52	8.55

FERRIS WHEEL RIDES

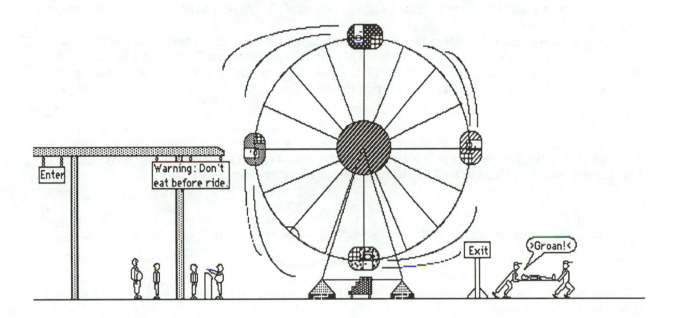

I. Wondering About...

...the forces that ferris wheel passengers feel as the wheel goes around.

II. Investigation to Undertake

Set up a spreadsheet to compute the forces that people feel as they ride around on a ferris wheel. What and where are the minimum and maximum forces that people on the ferris wheel feel? Draw a graph of the magnitude the force exerted on a passenger by the ferris wheel as a function of the angle of rotation from the top.

III. Background

Assume that the ferris wheel makes a complete revolution in about 10 sec, once it gets up to speed. Assume that it has a radius of about 6 m. Take the mass of a person to be about 75 kg.

IV. Related Science Concepts

This problem can be solved using Newton's second law of motion and some knowledge about motion in a circular path. For uniform circular motion,

$$a_c = r\,\omega^2,$$

13

where a_c is the centripetal acceleration, r is the radius of the object's circular path, and ω is the rotational velocity in radians per second. The centripetal acceleration always points toward the center of the circle.

From Newton's second law of motion;

$$a = \frac{\sum F}{m},$$

where m is the mass of the object and the acceleration of the object in this case is the centripetal acceleration.

Combining these two concepts yields the equations for the forces exerted by the ferris wheel on a rider. The rider feels two forces: the force of gravity pulling vertically downward and the force of the ferris wheel seat and constraints. These two forces must add up to a force pulling along the radius of the circle toward the center of the wheel.

EQUATIONS

Let ϕ be the clockwise angle from the top of the ferris wheel, and assume that the ferris wheel is rotating in a clockwise direction with the positive x axis to the right and the positive y axis up. The x component of the force of the ferris wheel on a rider is given by

$$F_x = -mr\,\omega^2 \sin\phi.$$

The y component of the force of the ferris wheel on a rider is given by

$$F_y = W - mr\,\omega^2 \cos\phi.$$

These equations can be used in the spreadsheet to compute the forces exerted by the ferris wheel on the rider every 15 deg.

ARROWS SHOT FROM A COMPOUND BOW

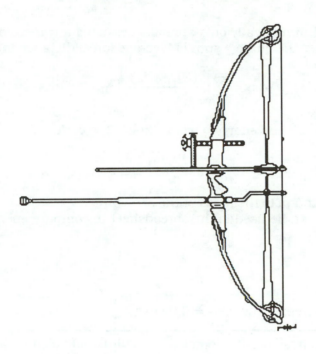

I. Wondering About...

...the speed of arrows after being shot from a compound bow.

II. Investigation to Undertake

Set up a spreadsheet to compute the energy and speed of an arrow shot from a compound bow as a function of the distance the bow string is pulled back. Mount a compound bow rigidly to a table and pull the string back a measured distance with a spring scale to determine the amount of force in newtons required to pull the bow back a given distance in centimeters. Take at least 25 data points. Then use the concepts of work, kinetic energy, and speed to complete the spreadsheet.

III. Background

Archery is a sport of increasing popularity for hunting and for target shooting. The compound bow is a relatively recent addition to the ancient use of bows and arrows. This investigation allows you to examine the physics of the compound bow. What are its advantages? Why was it invented?

IV. Related Science Concepts

The work done on the arrow by pulling it back against the force exerted by the bow string is equal to the energy of the arrow, according to the work-energy theorem. The work done will be equal to the area under the curve on a graph of the force versus the distance of pull.

EQUATIONS

This can be computed numerically on a spreadsheet using a trapezoid rule for the area:
    Work = area under the curve = sum of (average force)(distance increment),

$$\text{Work} = \sum_i \left\{ \left[\frac{(F_{i+1} + F_i)}{2} \right] [x_{i+1} - x_i] \right\}.$$

Since in this case kinetic energy = work ; $(mv^2)/2 = \text{work}$, so

$$v = \sqrt{\frac{2(\text{work})}{m}},$$

where v is the speed and m is the mass of the arrow.
    These equations can be used in the spreadsheet to compute energy and speed.

SAMPLE DATA:

COMPOUND BOW

Pull distance versus force; arrow mass = 0.023 kg

Pull Distance(cm)	Force(N)	Pull Distance(cm)	Force(N)
0.0	0	14.0	137
2.0	10	14.4	147
3.3	20	15.4	157
5.0	30	16.3	167
6.0	40	17.4	176
7.0	50	18.3	181
7.9	59	20.7	176
8.7	69	21.7	166
9.4	79	22.8	156
10.3	89	23.7	146
11.0	98	24.4	137
11.7	109	25.0	127
12.3	117	26.0	117
13.0	127		

Investigation 9

A JET AIRPLANE LANDING

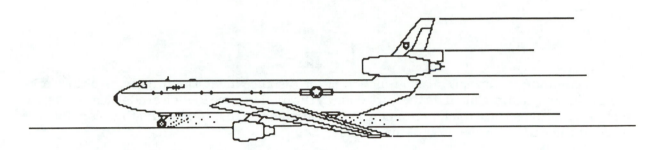

I. Wondering About...

...how the speed of a landing jet airplane changes as the plane brakes to a stop on the runway.

II. Investigation to Undertake

Set up a spreadsheet to compute the speed of the landing airplane as a function of the distance it has traveled along the runway. Display your results on speed versus distance graphs. Treat the landing speed at touchdown as an adjustable parameter. Examine the various landing speeds and stopping distances. Why might a beginning pilot be nervous if the landing speed is a bit higher than usual?

III. Background

Many inexperienced jet airplane pilots seem to have a tendency to panic if the landing is not going according to what they expect. Humans are linear thinking animals and it seems that pilots expect the airplane to have slowed down to half of its original landing speed after half of the runway has been used up. Is that really true? Use a spreadsheet to find out. A typical jet airplane trainer stops in about 4000 ft when it lands at a speed of 130 knots (1 knot = 1 nautical mile/ hr = 6000 ft/hr). Calculate the speed every 300 ft down the runway.

IV. Related Science Concepts

Brakes stop an airplane by converting its kinetic energy to heat energy because of the friction forces in the brakes. You can use the work-energy theorem from general physics to compute the results.

Change in kinetic energy = work done by the brakes.

Given: An airplane of mass M that lands at a speed of v_o can brake to a stop in a distance of d. Assume that the braking force F exerted to stop the airplace is constant, independent of airplane speed or distance of travel. (This is probably not quite true, since hot brakes tend to fade, but it should give about the right results.)

EQUATIONS

To compute the friction force F for a typical landing, touchdown speed v_o, and stopping distance d, use the work-energy theorem with the final speed set to zero:

$$\frac{Mv_o^2}{2} = Fd,$$

$$F = \frac{Mv_o^2}{2d}.$$

To compute the speed, v, as a function of the distance along the runway, use the work-energy theorem again. Calculate the work done by the friction force in each section (call its length s) of the runway, reduce the kinetic energy of the airplane by that much, and then compute the speed at the end of that section of the runway.

Final kinetic energy at end of section – initial kinetic energy at beginning of section
= work done by the brakes during that section of travel

$$\mathrm{KE}_f - \mathrm{KE}_i = \text{work done} = -Fs.$$

These calculations can be repeated until the kinetic energy, and hence the speed, are reduced to zero.

V. Mathematical Methods

A spreadsheet is an excellent tool to use to do repeated numerical calculations. For this investigation set up one column of your spreadsheet as the distance along the runway in 200 ft increments. Then each row can contain, for example, the distance, the kinetic energy at that distance as a fraction of its initial value, the speed at that distance, the percentage of the initial speed still remaining at that distance, and the time the airplane has been on the ground. Let r stand for the number of a row of the spreadsheet, then the row just above it will have the number $r-1$. The equations given in the section above can be combined with values in two adjoining rows of the spreadsheet to calculate the speed of the airplane at the end of each 200 ft increment of distance traveled down the runway.

$$\frac{1}{2}Mv_r^2 - \frac{1}{2}Mv_{r-1}^2 = -Fs.$$

Substitute an expression for F from the above equation for the friction force determined by the normal stopping distance d and then solve for v_r, the value of the speed in the rth row of the spreadsheet.

$$v_r = \sqrt{-\frac{v_o^2 s}{d} + v_{r-1}^2}.$$

This equation can be used in the spreadsheet to compute speed versus distance.

Since the force is assumed to be constant, we can calculate the time to cover each segment also:

$$t_r = \frac{2s}{v_r + v_{r-1}}.$$

This equation can be used in the spreadsheet to compute the time to travel the rth segment of runway. Accumulating the results gives the total time on the runway.

THE STOPPING DISTANCE OF CARS

I. Wondering About...

...how far does a car travel before it stops from the time the driver first realizes the need to stop.

II. Investigation to Undertake

Set up a spreadsheet to calculate the total distance a car travels from the recognition of the need to stop until the speed comes to zero including the reaction time of the driver. A reasonable reaction time is about 0.2 sec. Generate distance to stop versus initial speed graphs for several situations ranging between dry pavement and a more slippery pavement. How does the distance traveled while braking compare with the distance traveled while reacting (until the braking starts)?

III. Background

Accidents occur when drivers are unable to come to a stop before they meet some obstacle in the road. Often that obstacle is another car. If one drives at night with lights that illuminate the road for a few hundred feet, then one should not drive any faster than the car can be brought to a stop in that distance. If the road is very curvy and one can only see for a few hundred feet down the road at any one time, again one's speed should be low enough so that the stopping distance is less than that distance.

A knowledge of the stopping distances from various speeds on a car for various road conditions is important for safety. Of course, for maximum benefit, this knowledge should be coupled with the practical ability to estimate distances.

IV. Related Science Concepts

The ability of a car to control its motion depends on Newton's third law of motion: For every force exerted by one object on another, the second object exerts an equal and oppositely directed force back on the first. If a car exerts a force on the road, the road will exert an equal and oppositely directed force back on the car. This principle applies equally to propulsion, steering, and braking. Thus the ability of a car to stop quickly depends on the friction forces between its tires and the road. When tires are rolling, the tire surface is not sliding with respect to the road surface, so static friction determines the forces that are applied. If a force greater than that which is possible through static friction is applied, the tires will begin to slide on the pavement and kinetic friction takes over. Unfortunately, kinetic friction forces are usually smaller than static friction forces, thus it takes longer in time and distance for the car to come to a stop if static friction forces are exceeded.

The coefficient of static friction on dry pavement ranges from $\mu = 0.98$ for a very expensive sports car, which can also do 0 to 60 mph in 3.6 sec, to 0.63 for an economy model, which does 0 to 60 mph in 10.7 sec. The most slippery situation on the road is one in which the road is covered by thick, hard ice and the car has nonstudded tires. (In some states during limited times of the year it is permissible to use special snow tires that have metal studs installed.) In this worst case all cars have a coefficient of static friction of about $\mu = 0.1$.

EQUATIONS

The force that can be applied via friction is

$$F = \mu N,$$

where μ is the coefficient of friction and N is the normal force exerted between the vehicle and the surface that it is on. For a level surface, $N = mg$, where m is the mass of the vehicle and g is the acceleration due to gravity ($g = 32.2$ ft/sec^2).

With the force calculated from the above, use a technique similar to that used in the airplane landing investigation to generate your data.

Investigation 11

CENTER OF MASS

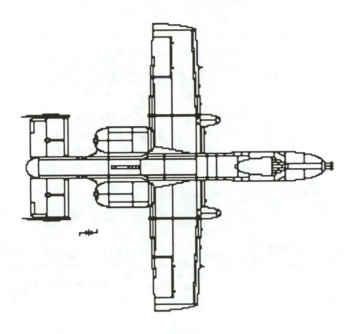

I. Wondering About...

...how to find the center of mass of a complex shape.

II. Investigation to Undertake

Set up your spreadsheet to imitate a complex, two-dimensional shape, putting numbers in the various cells to represent the mass of the various regions of the shape. Then use the functions of the spreadsheet to find the *x* and *y* (or column and row) centers of mass for the shape.

    You can do this by tracing the outline of the shape in which you are interested over a blank printout of your spreadsheet with the cells square. Then put into each cell the appropriate mass value for that region of the shape.

III. Background

An extended object has a point that behaves as if its whole mass were located there. The point, called the center of mass, is easily located experimentally by finding the one point where an object can be supported against the pull of gravity. To find the center of mass analytically is a straight forward calculus problem, usually treated in the first course in integral calculus, for objects whose borders are easily described by analytic functions.

    What about complex objects whose shapes do not lend themselves to analytic functional descriptions? The numerical methods of the spreadsheet can be used to solve such problems.

IV. Related Science Concepts

A complex object can be treated as a collection of particles, or on the spreadsheet, as a collection of cells. The center of mass of a system of particles depends only on the masses of the particles and the positions of the particles relative to one another. The coordinates of the center of mass of a collection of particles are given by the following equations:

EQUATIONS

$$x_{c.m.} = \frac{\sum m_i x_i}{\sum m_i} ; \quad y_{c.m.} = \frac{\sum m_i y_i}{\sum m_i} .$$

The column and row numbers can be used as the x and y coordinates, and the values in each cell can represent the mass of a part of the object at that location.

Use the sample data below as a practice set and determine the x and y (column and row numbers) coordinates for the center of mass of the object shown on it.

Some sample data for the aircraft pictured follow:

	A	B	C	D	E	F	G	H	I	J	K	L	M	N	O	P	Q	R	S	T	U	V	W
1										7	6	6	2										
2										10	10	10	4										
3										10	10	10	4										
4										10	10	10	4										
5										10	10	10	4										
6										10	10	10	4										
7										10	10	10	4										
8	8	15	15							15	15	15	10	5									
9	6	10	10			10	10	10	10	15	15	15	10	5									
10	6	10	10			20	20	20	20	10	10	10	8										
11	8	12	12	4	4	20	20	20	20	15	15	15	12	14	14	14	15	15	15	14	12	4	
12	18	18	20	20	20	20	22	22	24	24	26	26	26	28	28	28	30	30	30	28	24	12	6
13	8	12	12	4	4	20	20	20	20	15	15	15	12	14	14	14	15	15	15	14	12	4	
14	6	10	10			20	20	20	20	10	10	10	8										
15	6	10	10			10	10	10	10	15	15	15	10	5									
16	8	15	15							15	15	15	10	5									
17										10	10	10	4										
18										10	10	10	4										
19										10	10	10	4										
20										10	10	10	4										
21										10	10	10	4										
22										10	10	10	4										
23										7	6	6	2										

Investigation 12

LAUNCHING A ROCKET

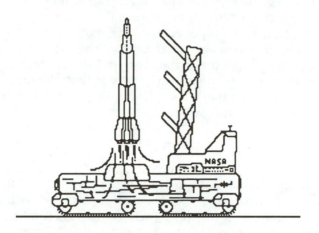

I. Wondering about...

...what it takes to launch a rocket.

II. Investigation to Undertake

Create a spreadsheet that simulates the launching of a rocket to an altitude of about 150 miles. (How far is that in kilometers?) The rocket's performance is a function of the velocity of the escaping gases from the rocket nozzle, the fraction of the rocket's mass exhausted per second, and the total time that the rocket engine burns. These should be variables with which you can experiment to accomplish the mission to launch a rocket to orbiting altitude. Remember, you must have some remaining rocket mass left when you get to orbit altitude.

If you are planning to carry out this and the next investigation, then you need to launch to orbit altitude using the minimum amount of rocket mass possible. You will start the next investigation with the quantities left at the end of this investigation.

III. Background

In the early part of this century, Robert Goddard was experimenting with rockets and dreaming of space flight. People wondered how a propulsion system would work in the vacuum of space. After all, there would be no air to push against. They did not realize, as Goddard did, that rockets do not push against the air. The rocket works on the principle of conservation of momentum. Just as tossing a weight out of a stationary boat causes the boat to start to move, the rocket pushes small amounts of gases out its nozzle at high velocities every second. Pushing on the gases causes a reaction force on the rocket. The result of pushing very hard on many gas molecules over a period of time can be a very large velocity for what is left in the rocket.

Why does it take so much effort to get such small amounts of payload up there? What sort of accelerations and speeds are involved? Which part of the trip takes the most energy: getting up there or settling into orbit? Why do they launch to the east and never to the west? Use this and the following investigation to answer these questions.

IV. Related Science Concepts

A launch into orbit seems to be a very complex undertaking. It can be understood by resolving it into two distinct processes related to components of the forces acting on the rocket due to the earth, just as Galileo resolved the components of the motion of a projectile near the earth's surface. First, the rocket has to thrust against the pull of the earth to "lob" itself up high enough to get into orbit. Second, it has to get up to speed so that the centripetal force required to keep it in a circular orbit is exactly equal to the pull of gravity at that altitude. This investigation deals with the launch-to-orbit-altitude phase. We will be dealing with the upward, outward, or radial part of the rocket's motion against the pull of gravity.

EQUATIONS

First, let us look into how the propulsion of the rocket works. Imagine a short time interval, Δt, during a "burn" of the rocket engine. At the beginning of this interval the total mass of the rocket plus its remaining fuel and payload is M and it is moving forward with a velocity, v. At the end of the interval, Δt, a small amount of burnt fuel whose mass is Δm has been expelled from the rocket and is now moving with a velocity, u. Meanwhile, due to pushing back on this gas the rocket, whose mass is now $(M - \Delta m)$, is moving forward at an increased speed, $v + \Delta v$. At the beginning of the interval the total momentum is

$$Mv.$$

At the end of the interval the total momentum is

$$(M - \Delta m)(v + \Delta v) + \Delta mu$$

If during this time interval the system is subjected to a net external force, F, then the impulse equation tells us that $F\Delta t$ equals the change in momentum of the system, the final momentum minus the initial momentum.

$$F\Delta t = [(M - \Delta m)(v + \Delta v) + \Delta mu] - Mv.$$

The velocity of the exhaust gases, u, can be written as the sum of the velocity of this material at the beginning of the interval, v, and their velocity relative to the rocket just after leaving the rocket, v_r. With this we can rewrite the impulse equation, collecting terms, neglecting very small terms like $\Delta m \Delta v$, and taking into account the directions of the vectors involved:

$$M\frac{\Delta v}{\Delta t} = v_r \frac{\Delta m}{\Delta t} - F.$$

This is the same as saying that at any instant in time, the net force on the rocket (left-hand side of the equation) is equal to the thrust (first term on the right-hand side) minus the external force (second term on the right-hand side), which we have assumed is working against the thrust of the rocket. We are going to ignore air resistance in this investigation, but since you have considered its effect in previous investigations, you are welcome to add this feature to this investigation. Do not forget that the air gets thinner as you go up. This should have an effect on the magnitude of the resistance as you go up.

TO LAUNCH THE ROCKET

From a vantage point away from the earth the motion would be in the radial direction (directly away from the center of the earth). It involves work against the pull of gravity.

$$F = G\frac{MM_e}{r^2}.$$

Note that the symbol, r, in the equation is measured from the center of the earth. By substituting for F and dividing through by M, the thrust equation for the lift phase can be written

$$\frac{\Delta v}{\Delta t} = \frac{v_r}{M}\frac{\Delta m}{\Delta t} - \frac{GM_e}{r^2}.$$

Set up a spreadsheet whose calculations depend on the variables that you can adjust: v_r, $\Delta m/M$, and the duration of the rocket burn. The first column should be the time incremented by Δt. You will find that it may be useful to have the time increment Δt as an input variable. The next should be the thrust acceleration (first term on the right-hand side). The next column should be the acceleration due to gravity at altitude (the second term on the right-hand side). The fourth column should be the actual acceleration (the difference between the previous two columns). Starting with zero upward velocity, the next column should calculate the velocity from the acceleration. From the velocity and starting at zero height, the next column should calculate the altitude. The last column should calculate the remaining mass in percentage of the original mass. Vary your inputs until you can get to the desired altitude at the apex of the path. Be careful not to cause the demise of your astronauts by subjecting them to too much acceleration.

USEFUL CONSTANTS:

Mass of the earth = 5.98×10^{24} kg
Mean radius of the earth = 6.37×10^6 m
Gravitational constant = 6.67×10^{-11} Nm²/kg²

INSERTING A ROCKET INTO ORBIT

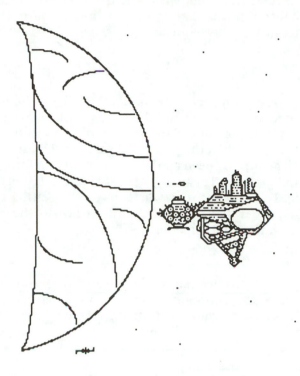

I. Wondering about...

...how to get a payload into orbit.

II. Investigation to Undertake

Create a spreadsheet that simulates getting the rocket from the previous investigation up to speed (tangent to the orbit) to stay in orbit. The rocket's performance is a function of the exhaust velocity of the escaping gases from the rocket nozzle, the fraction of the rocket's mass exhausted in each short time interval, the length of each short time interval, and the total time that the rocket engine burns. These should be variables with which you can experiment to accomplish each phase of the mission. In accomplishing the mission, you will have to determine the speed needed to sustain a circular orbit at the desired altitude. Remember, you must have some remaining rocket mass left when you get into orbit. You must be able to deliver some payload, say 5% of the original mass, into orbit or you have wasted your effort.

III. Background

Most of us have watched a rocket launch on television. Maybe it was the Shuttle or an earlier Apollo, Gemini, or Mercury launch. They seem to start moving straight up and then tip over into a curved path to merge into orbit. Those of you who have viewed the attempted launch of the fateful Challenger Shuttle in January of 1986 may recall that they do not actually start at full throttle on the Shuttle. They do not go to full throttle until 70 or 80 sec into the flight; after they have rolled over and ceased to be traveling straight up.

Why does it take so much effort to get such small amounts of payload up there? What sort of accelerations and speeds are involved? Which part of the trip takes the most energy: getting up there or settling into orbit? Why do they launch to the east and never to the west? Use this and the previous investigation to answer these questions.

IV. Related Science Concepts

A launch into orbit seems to be a very complex undertaking. It can be understood by resolving it into two distinct processes related to components of the forces acting on the rocket due to the earth, just as Galileo resolved the components of the motion of a projectile near the earth's surface. First, the rocket has to thrust against the pull of the earth to "lob" itself up high enough to get into orbit. Second, it has to get up to speed so that the centripetal force required to keep it in a circular orbit is exactly equal to the pull of gravity at that altitude. It has to be at this speed when it reaches the right altitude. This investigation will focus on the getting-up-to-orbit-speed phase of the mission. We will consider the rocket's operation not in the radial or upward direction, but in the horizontal or tangential direction.

EQUATIONS

Since we are not working against gravity in this part of the mission to get the payload into orbit, the external force is zero; therefore, the equations from the previous investigation must be modified to reflect this fact. The result is an equation that says the net force on the rocket is the thrust,

$$M\frac{\Delta v}{\Delta t} = v_r \frac{\Delta m}{\Delta t}.$$

We are going to ignore air resistance in this investigation. We are imagining an idealized mission in which not much tangential motion occurs before the rocket gets well above most of the air in the atmosphere.

THE ORBIT INSERTION PHASE OF THE MISSION

This part of the mission relates to the horizontal part of the trajectory. From a distance this is the tangential motion in the orbit. If we ignore air resistance, there are no external forces on the rocket in this direction. There are no tangential forces, but there is still the force due to gravity in the radial direction. For uniform circular motion the net force must be the centripetal force; therefore, we must be going at the right speed such that the centripetal force required is exactly equal to the force due to gravity at that altitude.

$$\sum F = M\frac{v^2}{r} = G\frac{MM_e}{r^2}$$

$$v = \sqrt{G\frac{M_e}{r}}$$

The orbital velocity that you need for insertion can be calculated from the altitude achieved in the lift portion of the mission. It is your goal in the insertion phase to achieve this velocity before you "use" the last 5% or so of your rocket's original mass. You may be surprised at the speeds that you have to achieve to get into orbit. It turns out that you can get a small boost from the rotation of the earth. From the launch site in Florida, you will have about a 380 m/s velocity in the tangential direction if you launch to the southeast. (You cannot launch due east because then you would not be orbiting the center of the earth. You must orbit in a great circle.) So you can start with this velocity instead of zero for this part of the mission. (Would there be any advantage to launching from Maine instead of Florida on this account?)

You could use a new section of the spreadsheet from the last investigation with a separate set of the same inputs and use the final results from the launch phase. Since now the external force is zero and we are really only interested in the velocity and remaining mass percentage, this spreadsheet need only have time, thrust acceleration, velocity, and mass percentage columns. Adjust the inputs until the necessary orbital velocity is achieved without using up all of the rocket's mass. Be careful not to cause the demise of your astronauts by subjecting them to too much acceleration.

MISSION DESCRIPTION

Having completed the two parts of a launch into orbit mission in these two investigations, you can put it all together in the form of a mission description such as: Burn for so many seconds from lift-off at a certain mass fraction per second. Wait until so many seconds after lift-off, then burn for so many additional seconds at a certain mass fraction per second. The time from lift-off to orbit insertion completion will be so many seconds. The deliverable payload will be such and such percentage of the original rocket mass. The orbit altitude will be so many meters above the surface of the earth.

How would your mission be described?

Investigation 14

SMALL CARS COLLIDING

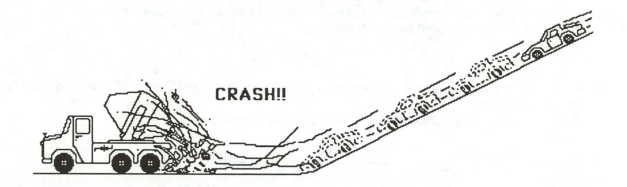

CRASH!!

I. Wondering About...

...why people in small cars are hurt worse in automobile accidents.

II. Investigation to Undertake

Set up a spreadsheet to compute the impulse on a small car when it collides with a larger vehicle. Assume the small car rolls down a hill from a fixed height and runs into a large stationary vehicle. Start with a small mass for the car and increase it systematically until it has a much larger mass than the stationary vehicle. How does the impulse imparted to the car change as the mass is increased?

III. Background

The fatality rate is greatest among passengers of small cars involved in accidents with larger vehicles. A very small car may have a mass of about 600 kg while a typical truck will have a mass 10 times that much.

If the car rolls down an incline from a vertical height of about 4 m, it will have a speed of about 9 m per sec (almost 29 mph) when it collides with the vehicle at the bottom. How will the impulse on the small car (and its occupants) change as the mass of the car is allowed to increase?

IV. Related Science Concepts

This investigation makes use of the conservation of linear momentum. The total momentum of both vehicles before and after the collision will remain constant. The conservation of energy will allow you to compute the velocity of the small car just before the collision if friction is neglected. In a collision of one car with another it is typical that about 20% of the initial kinetic energy is converted into deformation energy as the shapes of the colliding vehicles are changed.

31

EQUATIONS

Kinetic energy just before collision = initial potential energy,

$$\frac{mv_c^2}{2} = mgh,$$

where m is the mass of the small car, h is the starting height above the level street, g is the acceleration of gravity, 9.8 m/s², and v_c is the speed of the car just before collision.

Initial momentum = final momentum,

$$mv_c = mv_f + MV,$$

where M and V are the mass and final speed of the vehicle that was initially at rest, and v_f is the final speed of the car.

Kinetic energy after collision = 80% of the initial kinetic energy

$$\frac{mv_f^2}{2} + \frac{MV^2}{2} = 0.80\left(\frac{mv_c^2}{2}\right)$$

Impulse imparted to the small car,

$$\text{Impulse} = \text{change in momentum} = m(v_f - v_c).$$

These equations can be used in the spreadsheet to compute the impulse experienced by the small car.

Investigation 15

A BOUNCING GOLF BALL

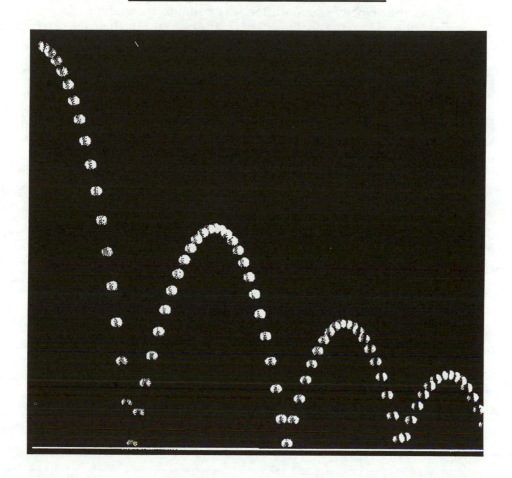

I. Wondering About...

...bouncing heights of a golf ball.

II. Investigation to Undertake

Set up a spreadsheet to analyze the bounce height of a golf ball. Collect numerical data from the picture above.

III. Background

A golf ball has a diameter of 2.5 cm. The strobe flashed at the rate of 30 times per second for the picture above. Set up a column for bounce number, bounce height before, bounce height after, velocity before, velocity after, and coefficient of restitution. Make a graph of the coefficient of restitution versus velocity before. Use additional columns to explore possible functional relationships between the coefficient of restitution and the velocity before.

IV. Related Science Concepts

Newton suggested that the velocity of separation was equal to a *constant* times the velocity of approach. This constant is called the coefficient of restitution, *e*. For the case of a bouncing golf ball Newton's suggestion can be written as

$$
\left\{ \begin{array}{c} \text{speed of rising} \\ \text{ball just after} \\ \text{leaving the floor} \end{array} \right\} = e \left\{ \begin{array}{c} \text{speed of falling} \\ \text{ball just before} \\ \text{hitting the floor} \end{array} \right\}
$$

How well does a constant value for *e* fit the filmed bounces of a golf ball? Try expanding the value for *e* as a power series in speed, that is,

$$
e = e_o + e_1 v + e_2 v^2 + \cdots,
$$

where the *e*'s are constants and *v* represents the speed of the ball.

Give a physical explanation for the best mathematical expression that you find for *e*.

THE ROTATIONAL SPEED OF COMPACT DISCS

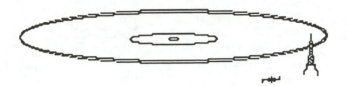

I. Wondering About...

...the distribution of information on a compact disc (CD).

II. Investigation to Undertake

Set up a spreadsheet to illustrate how dense the music data is packed in bits per radian from the inner edge of the data region of a compact disc out to the outer edge of the data region of the disc and how the rotational velocity varies. Do this for about 10 different radii.

III. Background

Compact disc players make music by converting a series of numbers to levels on a sound pressure wave. If these discrete values are converted frequently enough and the transitions between them are smoothed, a very remarkable recreation of the original sounds can be made covering the range of normal human hearing: 20 to 20,000 hertz or cycles per second.

The disc player accomplishes this feat by reading the numbers from the disc fast enough to store them in a computer-memory-type chip and then playing them out at a precisely controlled rate. The numbers are stored in binary form, that is, ones and zeroes corresponding to very small spots on the disc, which are detected by a tiny laser. Each one of these ones or zeros is called a bit, which is short for binary digit.

IV. Related Science Concepts

Compact discs rotate so that the linear speed of the disc where the data are being read is 1.25 m/s. The outer edge of the data region has a diameter of about 11.0 cm and the inner edge a diameter of about 4.50 cm.

The player reads bits in groups of 16 at a time to get each number, and the disc contains 44,100 of these numbers for each second of playing time. The maximum total playing time is 74 min and 33 sec.

EQUATIONS

The relationship between the angular velocity, ω, radius, r, and the linear or tangential velocity, v, is

$$v = r\,\omega .$$

DIVING FROM A HIGH BOARD

I. Wondering About...

...the rotation of a diver doing a competitive dive.

II. Investigation to Undertake

Set up your spreadsheet to use the conservation of angular momentum for divers after they have left the diving board, and also the ability of divers to change their body positions and hence the moment of inertia of the body, to compute the orientation of a diver in space as a function of time during a dive. Choose appropriate initial conditions and examine what changes in the moment of inertia must be made during a dive to have the diver go into the water head first with as little splash as possible.

III. Background

During the filming of the scenes for a physics videodisc, "Studies in Motion" (Great Plains National Television, 1983), film was taken of both male and female divers doing competitive dives. They leave the diving board giving themselves some angular motion by pushing off of the board with their feet while pulling the upper portions of their torsos forward. However, once they leave the board there is no longer any external torque acting upon them. All of their subsequent rotational motions are governed by the conservation of angular momentum and their ability to change their body configurations to control their orientation in space. To complete this exercise, compare your results either to a diving sequence on the above referenced videodisc or the line drawings of a diver provided on the next page.

IV. Related Science Concepts

The conservation laws are an important aspect of physics. When a system is not being acted upon by an external torque, then its angular momentum must remain constant. Once a diver leaves the diving board, the only force acting on the diver is the force of gravity. This force can be taken as acting vertically downward through the center of mass of the diver, so it exerts zero torque about the center of mass, and the angular momentum of a diver must remain constant. This fact can be expressed in the following equation:

EQUATIONS

$$\text{Angular momentum} = \text{constant} = I\omega,$$

where I is the momentum of inertia with units of kg m^2 and ω is the angular velocity in rad/sec.

The orientation of an object in space after a time Δt is computed from the original orientation in space by the equation:

$$\phi_2 = \phi_1 + \omega_1 \Delta t .$$

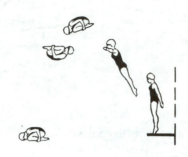

Investigation 18

FORCES IN LIMBS

I. Wondering About...

...the forces exerted in the bicep muscle in your arms and in the Achilles tendons in your legs during exercise.

II. Investigation to Undertake

Set up your spreadsheet to compute the forces in your biceps and your Achilles tendons for simple situations. Treat your arm as a third-class lever with the bicep applying a force between your elbow (pivot point) and your hand (the load). Treat your foot as a second-class lever with the ball of your foot (the pivot point) serving as contact point with the floor; your Achilles tendon applies the force that lifts your weight (the load) applied to the floor through your ankle. By varying the loads determine the forces that must be applied by your muscle or tendon.

III. Background

Biomechanics is a subject of considerable interest that makes use of the fundamental principles of physics to examine the physical properties of the human body. In this investigation you can use the conditions for equilibrium to determine the forces that must be exerted by some of the muscles and tendons in your body. You will then be able to understand how some of these muscles and tendons are readily injured in situations where excessive forces, such as in certain sporting events, are applied to parts of your body.

IV. Related Science Concepts

Take the pivot points for the motion of your arm or your foot as the center of rotation and set the sum of the torques about the pivot equal to zero. The resulting equations will enable you to compute the forces applied by your muscles or tendons as a function of the load you are attempting to overcome.

Take the necessary distance measurements from your body to estimate the distance from your elbow to where your biceps connect to your lower arm and the distance from your hand to your elbow. In a similar way measure the distance from your Achilles tendon to your ankle and from your ankle to the ball of your foot.

PULLING A CAR OUT OF THE MUD

I. Wondering About...

...the tension in a rope used to pull a car out of the mud.

II. Investigation to Undertake

Use a spreadsheet to calculate the tension in a rope as it is used to pull a car out of the mud. The rope is strung between a tree 10 m from the car and is pulled taut to a certain starting tension, say 200 lb. The rope is then pushed sideways at its center. Calculate the tension in the rope and the amount of force used to push the center of the rope as a function of angle between the halves of the rope and its original position. Express your results in SI units and make a graph.

III. Background

An old farmer's trick to pull a tractor out of the mud is the one described above. Very large tensions are developed in the rope. If the rope can survive the tensions that result, the tractor can be successfully extracted.

   Most ropes have a maximum elongation before they break. As the tension is increased, they stretch, never to relax quite back to their original length. They will stretch as much as 60% of their original length at their breaking point in the neighborhood of 5000 lb of force for a typical $1\frac{1}{2}$-in diameter climbing rope.

IV. Related Science Concepts

Assume that the principles of statics apply to the rope being held at any given deflection: The sum of the x components of the forces is zero and the sum of the y components is zero at all times.

EQUATIONS

   Assume that Hooke's law applies to the rope for all amounts of stretch up to the breaking point, then

$$F = k\Delta x,$$

where F is the force on the rope, k is called the force constant, and Δx is the amount of stretch. The force constant can be calculated under this assumption using the fact that when the force is zero the elongation is zero. (*Hint*: Sketch a graph of the force versus the elongation using this point and the maximum stretch data above.)

THE TACOMA NARROWS BRIDGE

I. Wondering About...

...the collapse of the Tacoma Narrows Bridge.

II. Investigation to Undertake

Investigate the overlap of the torsional and vertical vibrational modes for the original Tacoma Narrows Bridge. Use an approximation for the frequency distributions and estimate the fraction of energy pumped into the vertical oscillations that could couple to the torsional motion. Vary the construction parameters to see how you might have been able to save the bridge before it collapsed if you had been asked to help.

III. Background

On November 7, 1940, the $6.5 million bridge across the Tacoma Narrows collapsed into Puget Sound. This bridge failure was made famous by the film of the collapse taken by Barney Elliott of the Tacoma Camera Shop. Of course, the bridge had vibrated almost from the beginning of its construction. In fact, it had earned the nickname "Galloping Gertie" given by the Tacoma residents. However, up until the day of its collapse all of its oscillations had been in the vertical plane. It was only on the day of its collapse that the bridge began to oscillate wildly in torsional modes. You may want to watch the "Puzzle of the Tacoma Narrows Bridge Collapse" videodisc (John Wiley & Sons, Inc.).

IV. Related Science Concepts

Treat the Tacoma Narrows Bridge as an I-beam cross section with the bridge suspended by equivalent springs on each side. The equations of motion for both translation of the center of mass and rotation about the center of mass can be written down using Newton's second law.

EQUATIONS

Translational motion:

$$Ma = -k(y_2 + y_1)$$

where M is the mass per unit length of the bridge, k is the effective spring constant, y_2 and y_1 are the vertical displacements of the two sides of the bridge, and a is the vertical acceleration of the center of mass.

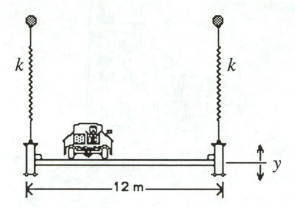

Rotation:

$$I\alpha = \text{net torque} = -kW\frac{(y_2 - y_1)}{2} \text{ for small angles,}$$

where θ is given by the small angle approximation

$$\theta \approx \frac{(y_2 - y_1)}{W},$$

I is the moment of inertia about the center of mass, and W is the width of the bridge.

V. Mathematical Methods

The solution of these two equations can be made by assuming that both y_1 and y_2 vary in the usual simple harmonic motion manner such that

$$y_1 = A_1\sin\omega_1 t \text{ and } y_2 = A_2\sin\omega_2 t.$$

Since these equations must be true for all times t, then it can be shown that there are two normal modes of motion and all real motion of this model of the bridge is a linear superposition of these two pure modes of vibration. These two modes are the pure vertical motion, where $A_1 = A_2$ and the frequency of vibration ω_v is given by

$$\omega_v = \sqrt{\frac{2k}{M}},$$

and the pure torsional motion where $A_1 = -A_2$ and the frequency of rotation ω_t is given by

$$\omega_t = \sqrt{\frac{kW^2}{2I}}.$$

For the original Tacoma Narrows Bridge the following numerical values can be deduced from the construction and behavior of the bridge:

$M = 4.4 \times 10^3$ kg per meter, $W = 12$ m , $R = 4.8$ m , $k = 1.5 \times 10^3$ N/m, and $I = MR^2$.

In a real system the normal mode frequency will not be a single frequency, but rather a distribution of frequencies. The energy per unit time that is accepted by a mode of oscillation is given by the following equation:

$$P(\omega) = \frac{1}{\left[(\omega - \omega_o)^2 + \left(\frac{\Delta \omega}{2} \right)^2 \right]} ,$$

where $\Delta \omega$ is the width of the resonance response curve at half maximum, which increases with the tendency of the system to resist oscillations. For the original Tacoma Narrows Bridge the tendency of the bridge to resist vertical and torsional motions was different. So the $P(\omega)$ function for vertical and torsional motion will have different values for both ω_o, the normal mode frequency, and for $\Delta \omega$. Use the constants given above to compute the ω_o values, ω_v and ω_t for vertical and torsional motion, and then use two different values for $\Delta \omega$ as adjustable parameters and vary them until half of the energy pumped into the vertical mode of oscillation by the wind gets transferred to the torsional mode of oscillation. Show a graph of $P(\omega)_v$ and $P(\omega)_t$ versus ω, shade in the overlap of the two curves.

Use your observations of the behavior of the bridge as you vary the $\Delta \omega$ values to suggest ways that the original bridge could have been saved.

THE SIZE OF GRANDFATHER CLOCKS

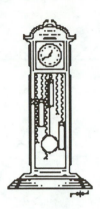

I. Wondering About...

...the size of grandfather clocks on various planets.

II. Investigation to Undertake

Set up a spreadsheet that computes the time for a simple pendulum to go from rest at one end of its swing to the other (one-half oscillation) at 1 deg intervals from 5 deg on one side of the vertical to 5 deg on the other. Determine the length needed for a 2-sec pendulum (1 sec for one-half oscillation) on the Earth, the Moon, Mars, Jupiter, and the Sun. Cells at the top should be reserved for you to input the length, acceleration due to gravity, and mass at the bottom of the pendulum. The spreadsheet should contain columns with the angular position, force, angular acceleration, angular velocity, and time. Experiment with different lengths until the right time is achieved.

III. Background

Some grandfather clocks are built so that each time the pendulum passes the vertical the hands are advanced by 1 sec. This requires that the time for one complete oscillation (from one side, across, and back) to be exactly 2 sec, since the pendulum passes the vertical twice in this motion. The time for one complete oscillation is called a period.

A pendulum is a nearly simple harmonic oscillator in which the restoring force is the force due to gravity. Since the force due to gravity on different planets is different, its effect on pendulums is different on each planet. How useful would a grandfather clock from Earth be to a settler on Mars or the Moon?

Typically, the pendulum is a rigid rod with a large mass at the end. The position of the mass can be adjusted by small amounts, which amounts to adjusting the length of the pendulum. The spreadsheet should be set up so that you can perform the equivalent action. Assume the clock pendulum is a simple pendulum.

IV. Related Science Concepts

Imagine the pendulum at an instant during its swing when it is a few degrees away from hanging straight down. The only two objects exerting forces on the mass at the end of the pendulum are the Earth (or planet that we are on) and the rod of the pendulum (whose mass we shall assume is negligible compared to the mass at the end). Establish a coordinate system with the y axis along the pendulum rod (perpendicular to the path of swing) and the x axis along the tangent to the swing as in the following diagram.

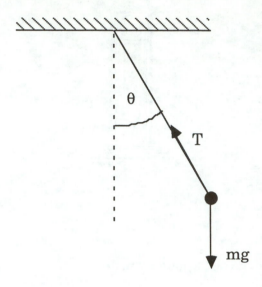

EQUATIONS

Since the mass at the end of the pendulum is not moving along the y axis, then the forces in this direction are in equilibrium and the tension in the string, T, is cancelling the y component of the force due to gravity, $mg \cos \theta$. This leaves only the x component of gravity, $mg \sin \theta$, as the net force on the pendulum.

$$\sum F = mg \sin \theta$$

This force is always oriented so that it is parallel to the tangent of the motion. You wish to derive the time, given the path of the object's motion, instead of wishing to derive the position as a function of time. The path will be a circular arc, equidistant on either side of the vertical. At each position of the pendulum, the force can be calculated. Since this force is tangent to the motion, you can write

$$\sum F = mg \sin \theta = ma_t = mr\alpha,$$

which allows you to compute the angular acceleration, α at each position. If the position intervals used are small enough, you can assume that α is constant over the interval. Using the equation

$$\omega_f^2 = \omega_o^2 + 2\alpha \, \Delta \theta$$

from angular kinematics, letting α be the average angular acceleration over the last interval and $\Delta \theta$ be the angular change during the interval, you can calculate the angular velocity at each position. Since

$$\omega = \frac{\Delta \theta}{\Delta t}$$

where ω is the average angular velocity over the interval, you can calculate the time at each position. The acceleration due to gravity of the Earth is $9.8\,\text{m/s}^2$; our Moon $1.6\,\text{m/s}^2$; Mars 3.8 m/s^2; Jupiter $25\,\text{m/s}^2$; and the Sun $270\,\text{m/s}^2$.

V. Mathematical Methods:

CONSEQUENCES OF TRUNCATION/APPROXIMATION ERRORS

You may find that even though the velocity should go from 0 at one end of the swing to 0 at the other end, the spreadsheet will either give a nonzero answer at the far end or it will give an error message at that cell for the velocity, even with the correct equation. Upon checking, you may find that the spreadsheet is trying to calculate the square root of a negative expression, because the terms (which should exactly equal each other) differ out beyond three or four decimal places. The solution to this problem is to round each term inside the square root expression off to about three decimal places while displaying two decimal places. Most spreadsheets have a ROUND (number, places) function. Use it on the last velocity cell. You could also use an IF (condition, then, else) function in the velocity column.

THE SIZE OF INTERVALS

The assumption of linearity made above is only good if the intervals taken are sufficiently small. Compare your results with those calculated using the period equation derived from the standard $\theta \approx \sin\theta$ approximation in your text. You will find a difference, even though we have stayed within 5 deg, which should result in a less than 0.2% discrepancy. Try two approaches to improve your results: (1) Keep the same number of intervals but make the maximum angle, say 1 deg and (2) double the number of intervals but stick to the same 5-deg maximum. Which gives the best results? In this context, what does it mean to "use small enough intervals"?

LIMITS ON THE ANGLE OF SWING

Your text gives the equation for the period of a pendulum derived under the assumption that the small angle approximation for $\sin\theta$ can be applied. The method that you have used in this investigation is not subject to this limitation. Can you extend your spreadsheet to determine where in angle of swing the small angle approximation breaks down? What is a useful criterion for deciding when the approximation breaks down? Can you calculate the period for angles of swing as high as 45°, 90°, or even 180°?

DAMPED OSCILLATIONS

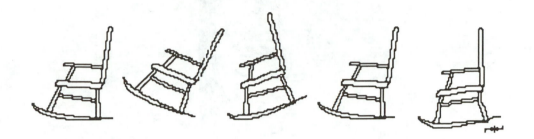

I. Wondering About...

...the back and forth motion of a rocking chair.

II. Investigation to Undertake

Set up your spreadsheet to enable you to compute the amplitude and period of oscillation for an oscillating system that loses energy. Try several different types of damping forces, for example, where the damping force is proportional to the distance travelled, or to the velocity, or to the square of the velocity. Which form of damping force seems to describe a system most like the ones you encounter in your everyday life? Why?

III. Background

In a general physics course, you generally study the idealized forms of simple harmonic motion, or at most, the case of damped harmonic motion where the damping term is proportional to the velocity of the moving object. Under these special conditions, the frequency of oscillation is not a function of the amplitude of oscillations for small amplitudes. It is a purpose of this activity to give you a chance to explore other types of damping forces.

IV. Related Science Concepts

Perhaps the most universally used force law in physics is the Hooke's law force. Hooke's law describes the behavior of a system that is subjected to a restoring force that is proportional to the displacement of the system from equilibrium and in the opposite direction to the displacement. Such idealized systems are conservative; that is the total energy is a constant.

However, in the real world most systems you will encounter are not conservative, but because of the presence of friction or drag or a dissipative force, the total useful energy of the system will decrease with time. This investigation will give you a chance to explore the behavior of such systems as a rocking chair or a swing.

51

EQUATIONS

Hooke's law: If a Hooke's-law-type force is the only force acting on an object, then

$$\sum F = ma = -kx,$$

where M is the mass of the system, a is the acceleration, x is the displacement from equilibrium, and k is the spring constant. This equation has the general solution for x as a function of time given by

$$x(t) = A\sin \omega t + B\cos \omega t,$$

where A and B are constants called amplitudes and ω is the frequency of oscillation in radians per second.

Dissipative forces can readily be added to a Hooke's law system for spreadsheet computations even though such forces can make analytic solutions to the Hooke's law equations quite difficult to obtain. The equation of motion for such dissipative systems can be written as follows:

$$Ma = -kx - F_d(x, v, v^2, \ldots),$$

where $F_d(x, v, v^2 \ldots)$ represents the dissipative force that may be a function of the displacement, and/or the velocity, and so on.

V. Mathematical Methods

The procedure to use to solve this type of problem, which may not lend itself to an exact analytical solution, is to set up your spreadsheet so as to use its iteration, or circular computation, capability. Imagine the motion of the rocking chair to be a specific example of a damped simple harmonic oscillator. The angular position of the rocker corresponds to the displacement of the oscillator, and the force tending to bring the rocker to equilibrium is the restoring force. Carry out all calculations for at least five complete oscillations. If you wish, you can try three different functional forms for the dissipative force; for example, proportional to the total distance traveled, or to the velocity, or the velocity squared. By using the graphing part of your spreadsheet, you can see the effect of such forces on the amplitude and the frequency of oscillations.

SINGING RODS

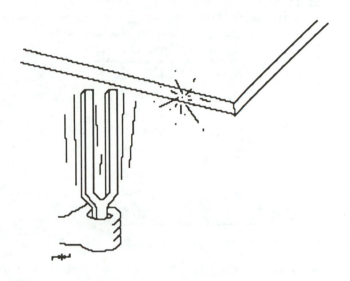

I. Wondering About...

...sound waves resonating in metal rods.

II. Investigation to Undertake

Set up a spreadsheet to calculate the resonant lengths of rods of various metals to generate a musical scale. The resonances that make the notes are excited by clamping the rods at the midpoint and generating longitudinal vibrations in the rods by tapping their ends or rubbing them lengthwise using some rosin for friction. Your output should include wave speed, frequency, wavelength, and rod length for each note. These numbers should be generated based on the bulk modulus of elasticity, the density of the metals involved, and the set of frequencies for the musical scale given below.

III. Background

An exceptionally loud and persistent ringing can be generated by exciting metal rods into longitudinal vibrations. Rods or tubes will work. Because the vibrations are longitudinal, the diameter of the rod does not matter, but vibrations are easier to excite in rods of 1.0-cm diameter or less. They should be cut reasonably square and can be made of about any metal found in a shop. To get some stick-slip friction on the rod, rub it with a cake of rosin to distribute it along the length. Grasp the rod at its center with the thumb and forefinger of one hand and with the thumb and forefinger of the other hand stroke the rod. With a little practice one can get the rod to "sing" in much the same way as a wine glass can "sing" when one rubs its edge.

IV. Related Science Concepts

These resonant systems vibrate with a standing wave pattern consisting of a fixed set of nodes and antinodes. For a vibrating rod clamped at its center, the pattern consists of antinodes at the ends of the rod and a node at the center. Since the distance between two nodes or two antinodes is equal to one half of the wavelength of the fundamental, then the length of the rod in each case is one half the wavelength desired.

The frequencies desired are those of a standard (equal-tempered) musical scale as follows.

Note	Frequency (Hz)
A	1760.0
B	1975.5
C	2093.0
D	2349.3
E	2637.0
F	2793.8
G	3136.0
A	3520.0

EQUATIONS

The wave speed, v, wavelength, λ, and frequency, ν, of a wave are related by the equation

$$v = \lambda \nu .$$

The speed of a sound wave in a solid is

$$v = \sqrt{\frac{B}{\rho}},$$

where B is the bulk modulus of elasticity of the solid and ρ is the density of the solid. The bulk modulus and density of various typical metals are as follows.

	Density (10^3 kg/m³)	Bulk Modulus (10^9 N/m²)
Aluminum	2.70	70
Brass	8.50	61
Copper	8.93	140
Steel	7.82	160

DOPPLER SHIFT

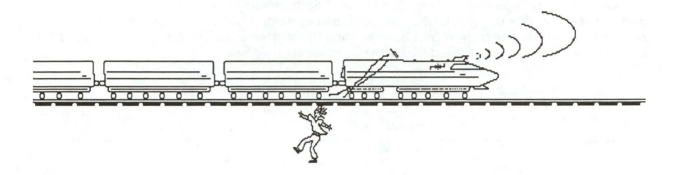

I. Wondering About...

...how the frequency of a siren changes as it goes by you.

II. Investigation to Undertake

Generate a spreadsheet that calculates the frequency of a sound wave coming toward you as a function of time as the source of the sound passes by. Assume that you are standing a distance, d, away from the straight line path traveled by the emitter of the sound. When it is directly opposite you on the path, it is at an angular position of 0°. Positive angles are for positions before that point and negative after. Allow the speed of the emitter, v_s, the distance you are standing from the line of travel, d, and the frequency of the emitted sound, v_s to be variables that can be changed in input cells. Allow the angular position of the source to range from +80° to -80° in steps of 10°. At each position calculate the observed frequency, the elapsed time (starting at 0 for +80°), and the position of the source on its line of travel. For position along the line of travel, call the point directly opposite the observer position 0 and measure in meters from there, using negative positions before this point and positive after.

Experiment with various values of the inputs, v_s, d, and v_s. Make plots of the observed frequency *versus* angular position, time, and position of the source along its line of travel.

Another problem you may want to explore is that it takes a finite amount of time for the sound to travel from the source to the observer so that what the observer hears is delayed. What is the frequency observed as a function of the current position of the source, instead of where it was when it emitted the sound?

III. Background

You have heard a siren of an ambulance or police car as it passes by or the horn of a car honking as it passes. There is a high pitch as it approaches and a lower one as it recedes and a shifting pitch as it is passing by. Your text gives an equation for the Doppler shift of sound for a source traveling directly toward an observer and directly away as if the observer were on the line of travel, but when we observe the Doppler shift of sound we never stand literally in the way of the moving source. You are always to the side. In addition, the equations in your text never attend to the change in frequency as the source passes by. It is this change that you are looking at in this investigation.

IV. Related Science Concepts

The Doppler equations in your text refer to the shift in frequency of sound between the source and the observer due to the relative velocity of the two along a line directly between them. If the source is not moving directly at the observer, then the component of the relative speed along the line between them must be used. Refer to the following diagram.

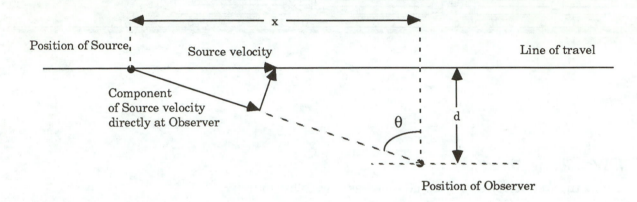

EQUATIONS

Trigonometry tells us the distance, x, is $d\tan\theta$, the distance between the source and the observer is $d/\cos\theta$, and the component of the source velocity directly toward the observer is the source velocity times $\cos(90° - \theta)$. Together with the Doppler shift equation from your text, you can use these relationships to build the desired spreadsheet and plot the results.

HYDRAULIC BRAKES OF A CAR

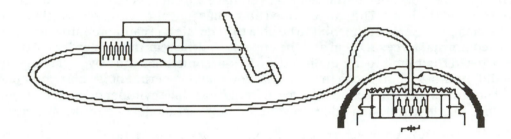

I. Wondering About...

...how the brake system in a car works.

II. Investigation to Undertake

Build a spreadsheet that models the brake system in a car. There should be separate columns for the force from your foot, the brake pedal lever arm length, the master cylinder lever arm, the force at the master cylinder, the diameter of the master cylinder, the pressure developed in the brake system, the radius of the slave cylinders, and the force developed at the master cylinders. Start with the specifications of an actual system below and vary some of them to discover the effect of the changes in subsequent rows of the spreadsheet. Is the force multiplied through both the brake pedal lever and the hydraulic parts of the system? If you find a place where it is not multiplied, why wouldn't the force be multiplied throughout the system?

III. Background

The brake system of a car is an often ignored performance component of a car; that is, until it fails. Although 0 to 60 or 100 mph times are often boasted about, few people know the 100 or 60 mph to 0 times for their cars and few realize that the deceleration times are shorter than the acceleration times.

The brake system of a car is activated by a force applied to the brake pedal. Your foot applies a force on a lever, pivoted at one end, about 20.0 cm from the pivoted end. A rod attached to the lever at about 7.5 cm from the pivot transmits a force from the lever to the master cylinder. The master cylinder has a diameter of 1.9 cm, and a pressure is generated in the brake fluid by the force delivered to it from the brake pedal. Brake fluid is an incompressible fluid that has a high boiling point and will not cause rust in the metal components of the system. The pressure is delivered through rigid metal tubes to the slave cylinders at each wheel. (There are some flexible plastic sections that are reinforced so that they do not expand with pressure.) The slave cylinders are about 3.8 cm in diameter each. This pressure is delivered to the brake pads at a distance of about 10.0 cm from the axis of rotation of the wheels. The brake pads themselves have a rubbing area of about 90 cm² each. The result of all of this is that this car whose mass is about 1.0 metric tons can stop with a 0.5-g deceleration when a force is applied to the brake pedal of 89 N, about 20 lb.

Since your stopping distance is limited by the static friction that can be developed between the tires and the ground, a practical limit is placed on the forces developed in the brake system. If this level is exceeded, then the friction between the tires and road will go kinetic and drop, causing your stopping distance and time to go up.

There are two other features that you *may* wish to explore:

(1) When you brake, because of the fact that you have a front and rear axle, the car's weight is shifted toward the front. The car described above has just about an even weight distribution. This causes the wheels in the front to be able to generate a greater stopping force, due to the increased normal force on them and the opposite for the rear tires. In order for the rear tires not to exceed their limit to maintain static friction, sometimes the slave cylinders in the rear are a different size than those in the front. You could experiment with various percentage weight shifts and determine the changes in front/rear slave cylinder diameters. If you choose to explore this feature, then add it to your spreadsheet model of a brake system.

(2) The kinetic energy of your car is dissipated as heat by kinetic friction between the brake pads and the brake disks (assuming four-wheel disk brakes). If too much heat is generated, the material that the pads are made of will form a vapor layer between the pad and the disk, which can cause the force delivered to the ground to be nearly zero. This situation is the extreme result of what is known as brake fade. If you choose to explore this feature, do some research to discover the parameters that control this situation and add them to your spreadsheet model of a brake system.

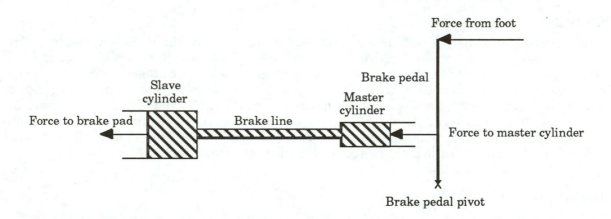

Schematic diagram of a brake system.

IV. Related Science Concepts

The brake pedal is a lever that is stationary (more or less) when one achieves the required braking results. The sum of the torques on this lever is zero under these conditions.

The hydraulic system works under the principle of hydrostatics known as Pascal's law, which states that for an incompressible fluid in a rigid closed system, the pressure is the same everywhere.

EQUATIONS

The torque is given by

$$\mathbf{\Gamma} = \mathbf{r} \times \mathbf{F},$$

where $\mathbf{\Gamma}$ is the torque, $\mathbf{r}$ is the lever arm, and $\mathbf{F}$ is the force applied.

Pressure, P, and force, F, are related by the expression

$$P = \frac{F}{A},$$

where A is the area over which the force is applied.

MAGDEBURG HEMISPHERES

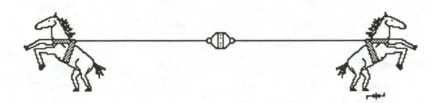

I. Wondering About...

...the forces exerted on the Magdeburg hemispheres.

II. Investigation to Undertake

Set up your spreadsheet to compute the total force exerted by the atmosphere on the Magdeburg hemispheres as a function of the radius of the hemispheres and the altitude of Magdeburg.

III. Background

In 1654 Otto von Guericke, burgomeister of Magdeburg and inventor of the air pump, gave a demonstration before the Imperial Diet in which two teams of eight horses could not pull apart two evacuated brass hemispheres. What is a minimum size that the hemispheres could have been? What is the maximum altitude at which Magdeburg could be located? Make a reasonable guess about the force a horse can exert. (Hint: 1hp = 746 watts. A draft horse can pull a heavy load at a constant speed of about 2/3 m/s.)

IV. Related Science Concepts

The pressure of the atmosphere acts equally in all directions on the external surface of the evacuated hemispheres. Hence, when the hemispheres are placed together and the air is removed from the interior volume, the force holding the two hemispheres together is given by the following equation.

EQUATIONS

$$F = \pi R^2 \Delta P,$$

where R is the radius of the hemispheres and ΔP is the difference between the interior and exterior pressure. Assume that von Guericke's air pump was able to remove 95% of the original volume of air inside the hemispheres; then ΔP is equal to 0.95 times atmospheric pressure at Magdeburg. Given below are some altitudes and typical atmospheric pressures for some locations of the world. Notice that the pressure *versus* altitude is approximately exponential.

Location	Elevation (m)	Pressure (atm)
Boston, MA	~ 0	1.000
Rapid City, S.D.	1000	
Gallup, N.M.	2000	
Mt. St. Helens, WA	3000	0.693
Mt. Zupo, Switzerland	4000	
Mt. Blackburn, Alaska	5000	
Mt. Logan, Canada	6000	0.435
Mt. Macha Pucchare, Nepal	7000	
Mt. Gosainthan, Tibet	8000	
Mt. Everest, Nepal-Tibet	8850	
	9000	0.292

LEAKY WATER CYLINDER

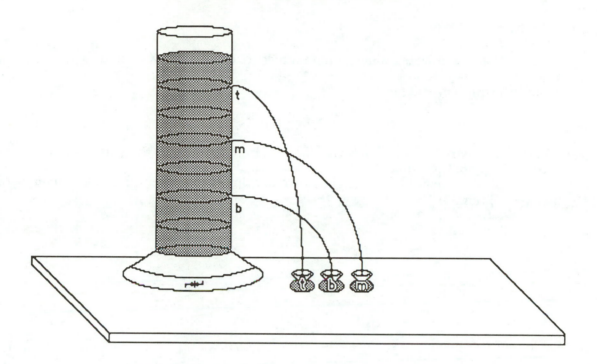

I. Wondering About...

...water running out the holes in the side of a cylinder.

II. Investigation to Undertake

Use the spreadsheet to investigate the flow of water out of a water tank cylinder that has three holes in its side. The three holes are *equally spaced* between the bottom and the top of the cylinder. Compute the flow of water out of each hole as a function of time. Compute the horizontal distance the water shoots out of each hole before hitting the table as a function of time. Plot the ranges of flow from the three holes on a graph and indicate the times when any two of the three streams of water land at the same location on the table.

III. Background

The water flowing out of the top hole (*t*) has the smallest horizontal velocity, but it has the greatest amount of time to fall before hitting the table. The water flowing out of the bottom hole (*b*) has the greatest horizontal velocity as it leaves the cylinder, but has the least distance and time to fall before hitting the table. The water coming out of the middle hole (*m*) has medium speed and medium time. How does the horizontal range of each stream of water compare as time passes?

IV. Related Science Concepts

The flow of water out of a hole will depend upon the water pressure at that level in the liquid. As the level changes so will the flow rate. Use the continuity equation and Bernoulli's equation to examine the behavior of this system.

EQUATIONS

The continuity equation is

Volume of water coming out of a hole in 1 sec = area X velocity.

Bernoulli's equation states

$$\rho g h = \left(\tfrac{1}{2}\right)\rho v^2,$$

where ρ is the density of the water, h is the distance from the hole to the surface of the water, and v is the velocity of the water.

Take as typical values: cylinder height = 46 cm; cylinder diameter = 5.0 cm; and diameter of the holes = 1.5 mm.

Investigation 28

HEART ATTACK RISKS

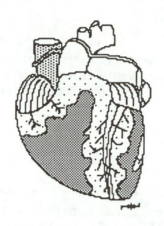

I. Wondering About...

...how life styles and age are related to heart attack risk.

II. Investigation to Undertake

Set up a spreadsheet to collect life style information and then compute a Heart Attack Risk Total (**HART**). Assume a linear reduction in the diameter of the coronary artery as a function of age and use the physics of laminar fluid flow to calculate the relative blood flow rate.

III. Background

There are nine life style and hereditary factors that have been identified as contributors to heart attacks. Set up a column of the spreadsheet into which you enter your appropriate life style or hereditary factor coefficient value as given below:

Factor	Coefficient Values
Parents	**0**-no heart death < 50 yr. old; **1**-one parent heart death < 50 yr. old; **2**- for both parents heart deaths < 50 yr. old.
Smoking	**0**-do not smoke; **1**-for smoking.
Cholesterol	**1**-normal or below; **2**-above normal; **1**-don't know.
Stress	**1**-normal; **2**-above normal; **3**-much above normal.
Weight	**0**-normal; **1**-slightly overweight; **2**-considerably overweight.
Exercise	**0**-frequent, aerobic; **1**-regular, periodic; **2**-little.
Personality Type	**1**-Type A (striving achiever, counter); **0**-Type B(even-tempered).
Diabetes	**0**-no; **1**-yes.
Hypertension	**0**-no; **1**-yes.

The total for your heart attack risk (**HART**) is computed from your life style and hereditary factor coefficient values by the following formula:

HART =   5*(Parents) + 9*(Smoking) + 4.5*(Cholesterol) + 1.5*(Exercise) + 1.5*(Weight) +
1.7*(Stress) + 3*(Personality Type) + 8*(Diabetes) + 3*(Hypertension).

IV. Related Science Concepts

ARTERIAL DIAMETER REDUCTION

Set up a column of your spreadsheet to compute the diameter of your coronary artery as a function of age beginning with your twentieth birthday. Assume a linear reduction in your arterial diameter with age, beginning with a 4-mm diameter at age 20 yr.

EQUATIONS

$$\text{Artery Diameter}(mm) = 4.0\left(1 - 6.0 \times 10^{-4}(\text{HART})(\text{time in years since } 20)\right)$$

IDEALIZED LAMINAR FLOW IN A RIGID TUBE

Set up a column of your spreadsheet to compute the idealized laminar flow of blood in your coronary artery based upon your **HART** and your age. Poisseuille's law gives the relationship between the diameter of a tube and idealized laminar flow:

$$\text{Rate of flow} = \left(\frac{\pi(P_1 - P_2)R^4}{8\eta L}\right),$$

where R is the radius of the tube, $(P_1 - P_2)$ is the pressure difference over a distance L along the tube, and η is the viscosity of the fluid flowing through the tube.

Use this equation to compute the relative blood flow as a function of age, that is, take 20 yr. as 100%.

REALISTIC FLOW FOR BLOOD IN A FLEXIBLE TUBE

Set up a column of your spreadsheet to compute a realistic value for the relative amount of blood flow in your coronary artery based upon experiments with human subjects. The variable viscosity of the blood and the flexible walls of the artery cause the idealized flow not to be a good representation of real flow. However, the logarithm of the ratio of the real flow to the idealized flow is well described by a simple polynomial in the reduction of the arterial diameter as follows:

$$\log_e\left(\frac{\text{real flow}}{\text{ideal flow}}\right) = 0.96(\Delta d) + 0.18(\Delta d)^2,$$

where Δd is the amount the diameter is reduced.

CRITICAL STENOSIS

Set up a column of your spreadsheet to compute your age for critical stenosis. This is the age at which your real coronary arterial blood flow is reduced by 7% in the realistic blood flow column; that is, it is 93% of your blood flow at age 20. This is the age at which your coronary risk is substantially increased.

GRAPHICAL RESULTS

Convert all of your blood flow data to a graph with your age on the horizontal axis and the relative blood flows on the y axis. Put a Critical Stenosis mark on your graph.

Investigation 29

COFFEE COOLING IN A CUP

I. Wondering About...

...the cooling rate of coffee in a cup.

II. Investigation to Undertake

Place a small amount of hot (80°C) water in a cup and measure its temperature every 20 sec or so. Set up a spreadsheet to calculate the cooling rate of the water as a function of time. Determine the mathematical expression between the cooling rate and time. Then use this relationship to investigate the question: If you are going to add a small amount of cream, say about 1/10 of the amount of coffee, at room temperature (20°C) to the coffee, but you want the coffee to be as hot as possible when you drink it in 3 min, should you add the cream immediately or wait until just before you are going to drink the mixture? Use your spreadsheet analysis to justify your answer.

III. Background

A typical cup of coffee holds about 8 oz (224 ml) of coffee. A small cream container holds about 23 ml.

IV. Related Science Concepts

Set up a spreadsheet to calculate such physical attributes of the cooling liquid as the amount of heat loss per unit time, the amount of temperature change per unit time, the fractional change in temperature per unit time, and so on, to see if you can find a functional relationship between some property of the system and the time of cooling.

Newton proposed a simple rule for the cooling of systems by convection. He suggested that the rate of cooling was proportional to the difference between the temperature of the object and

67

the temperature of the surroundings. Test your data against Newton's proposal. See the following sample data.

Cooling Rate Calculations from Experimental Data

Start Time =   0 sec   liquid volume =   224 ml

Interval =   10 sec   cream volume =   23 ml

room temperature =   20 °C

Time (s)	Temp (°C)
0	80.0
10	68.0
20	58.4
30	50.7
40	44.6
50	39.7
60	35.7
70	32.6
80	30.1
90	28.1
100	26.4
110	25.2
120	24.1
130	23.3
140	22.6
150	22.1
160	21.7
170	21.4
180	21.1

Investigation 30

THE SURFACE TEMPERATURE OF THE MOON

I. Wondering About...

...how the temperature of the moon's surface changes with time as it rotates out of the sun's light.

II. Investigation to Undertake

Set up a spreadsheet to calculate the temperature of the lunar surface as a function of the angle between directly overhead and the location of the sun in the lunar sky. Make a graph of the temperature as a function of angle. *Note:* 90 deg and 270 deg will be lunar sunset and sunrise, respectively.

III. Background

The moon rotates 360 deg in 29.5 earth days. The solar radiation falling on the moon's surface is 1.35×10^3 W/m². The reflectance of the moon is 0.076. The Stefan-Boltzmann constant is 5.67×10^{-8} W/m²K⁴. The maximum temperature of the surface of the moon facing the sun has been measured as 389 K. The minimum temperature for the dark side of the moon is 122 K.

IV. Related Science Concepts

Consider the conservation of energy for a thin unit area of the moon's surface. The energy input will include the energy from the sun and the energy from the moon's interior. The energy output

69

will be thermal radiation. The internal energy change will be the thermal capacity of the surface times the change in temperature.

EQUATIONS

$$\text{Solar energy absorbed per unit time} = S(1-r)\cos\left(\frac{2\pi t}{P}\right),$$

where t is the time from overhead, S is the solar radiation falling on the surface, r is the reflectance of the surface, and P is the rotational period of the moon.

$$\text{Rate of radiated energy}\left(\frac{\text{W}}{\text{m}^2}\right) = \sigma T^4,$$

where T is the temperature in Kelvin and σ is the Stefan-Boltzmann constant.

The rate of temperature change of the unit area of the lunar surface will depend upon its thermal capacity. *Hint:* Set up your spreadsheet with the high temperature, low temperature, and thermal capacity as adjustable parameters.

Investigation 31

HOT RODS

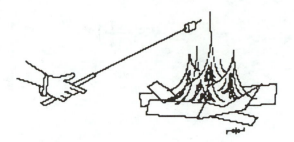

I. Wondering About...

...how thermal energy travels through rods of various materials.

II. Investigation to Undertake

Set up a spreadsheet to calculate the flow of thermal energy along several rods of different materials. Select a convenient row near the top of your spreadsheet and assume that it is a high-temperature thermal energy reservoir of constant temperature. Use a column of the spreadsheet below the reservoir to represent a rod of material of unit cross-sectional area, say 1 cm². Use each row along the column to represent a length, say 1 cm, along the rod. Put your spreadsheet on manual calculate. Put the proper equations in each cell along your rod. Then, using manual recalculations, watch the thermal energy flow down the rods. Use graphics to show the temperature distribution along a rod as time passes. Also compare the temperature distributions for different materials. Then compare the flow of thermal energy through a typical pane of glass to the flow through a double-pane window with an air space between the panes.

III. Background

The thermal conductivity of a material is a measure of its rate of thermal energy flow. The thermal conductivity of a material has the units of joules per second meter °C. Typical values are as follows.

Material	Thermal Conductivity(J/s m °C)
Air(STP)	0.026
Aluminum	237.
Construction brick	0.63
Copper	401.
Glass	0.8
Insulating block	0.147
Iron	80.
Oak	0.15

71

IV. Related Science Concepts

The rate of the flow of thermal energy by conduction is proportional to the temperature gradient. Thus the thermal energy flow Θ across an area A, in a material with a thermal conductivity K, where the temperature gradient is $\Delta T/\Delta x$ is given by the following equation:

EQUATION

$$\Theta = -KA\left(\frac{\Delta T}{\Delta x}\right).$$

MIXING HOT AND COLD LIQUIDS

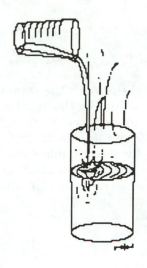

I. Wondering About...

...the final temperatures of various mixtures of hot and cold liquids.

II. Investigation to Undertake

Set up your spreadsheet to calculate the final temperature for various mixtures of hot and cold liquids. Use the graphics to produce a graph of the final temperature as a function of the amount of the second liquid added to the first.

III. Background

From a very early age we experiment with mixtures of liquids, from bath water to milk on breakfast cereals. There is something fascinating about bringing together two different liquids and stirring them into a homogeneous result. The purpose of this investigation is to use the concept of the conservation of energy to examine the final temperatures of various mixtures of liquids.

IV. Related Science Concepts

A method to determine the final temperature of a mixture of two liquids, originally at different temperatures, can be derived from the first law of thermodynamics. According to this law, for an isolated system the heat energy lost by one liquid will have to be equal to the heat energy gained by the other liquid. In other words, the total heat energy of an isolated system will remain constant.

Let the first liquid be the liquid at lower temperature so that it will gain heat energy from the second. Let M_1 stand for the mass of the first liquid, T_1 stand for its original temperature, and c_1 stand for its specific heat; also for the second, or higher temperature, liquid. Let T_f stand for the final temperature of the resulting mixture.

EQUATIONS

Heat Energy Gained by object 1 = Heat Energy Lost by object 2.

$$M_1 c_1 (T_f - T_1) = M_2 c_2 (T_2 - T_f).$$

You can solve this equation for T_f and then use the ratio of M_2/M_1 as a variable parameter of the system. After picking appropriate values for the starting temperatures of the two liquids, you can examine the behavior of the system as you change the relative amounts of the two liquids. If you wish, you can give the two different liquids different values of specific heat and see how that changes the results you obtain for a two-liquid system.

Use a graphics routine to plot the final temperature as a function of the ratio of the masses. What do you expect for the extreme mass ratios of zero and infinity? Do your numerical results agree with these expectations?

TIRE PRESSURES ON A LONG TRIP

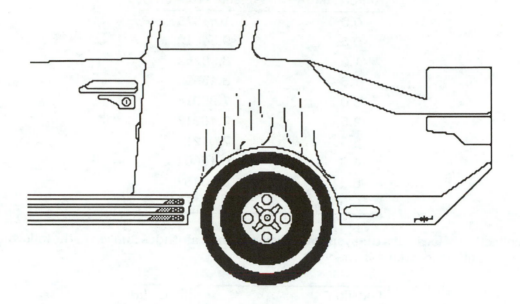

I. Wondering About...

...factors which influence pressure changes in automobile tires.

II. Investigation to Undertake

Develop a spreadsheet which illustrates the variations in pressure in a typical car tire due to changes in temperature and atmospheric pressure. Assume that you have filled the tire to a pressure of 29 psi at 20°C when the atmospheric pressure is 14.7 psi. Imagine that you live on a coast at nearly sea level and you take a trip to various cities at higher altitudes and that your tires warm up at times during the trip. Your spreadsheet should show the resulting tire pressure for a range of temperatures at four or five altitudes from sea level to the highest altitude city. Let the temperature range from 10°C to 50°C. (Have your spreadsheet convert all quantities to the proper SI units.)

III. Background

Car and tire manufacturers give recommendations as to proper tire pressures for optimum safety, handling, and even economy (fuel and tire wear) characteristics. These specifications are pressures to which tires should be inflated when they are "cold," nominally at about 20°C. As you drive and the tire temperature goes up, what does the tire pressure do? What is the effect of changing altitude? If you had forgotten to check your tire pressure before starting, but you do have a way of determining atmospheric pressure and tire temperature, how could you get the right amount of air in your tires?

IV. Related Science Concepts

Tire pressures given are stated in what is called gauge pressure, that is, the pressure above ambient atmospheric pressure. A table of standard atmospheric pressure follows.

Altitude (km)	Pressure (10^4 N/m^2)
0.0	10.1325
0.5	9.54612
1.0	8.98762
1.5	8.45596
2.0	7.95014
2.5	7.46917
3.0	7.01211
4.0	6.16604
5.0	5.40482
6.0	4.72176

A collection of locations that you could visit and their altitudes is given in the following table. You can pick other locations if you wish.

Location	Altitude (ft)
Boise, ID	2,880
Chicago, IL	580
Colorado Springs, CO	6,072
Denver, CO	5,279
Flagstaff, AZ	6,907
Los Angeles, CA	286
Pikes Peak, CO	14,108
San Francisco, CA	50
Santa Fe, NM	7,013
Snoqualamie Pass, WA	3,131
Yellowstone Park, WY	7,741

Assume that the air in the tires behaves as an ideal gas in a container of constant volume.

ATMOSPHERIC PRESSURE

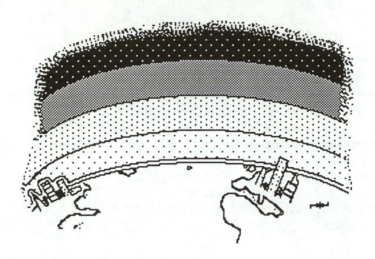

I. Wondering About...

...how the atmospheric pressure changes with altitude.

II. Investigation to Undertake

Given below are typical values for the pressure and density of the air above the earth. Try some different models to see how well you can approximate the dependence of the density of the air on altitude. Consider a constant density model, an isothermal model, and an adiabatic model.

III. Background

In the following table are given the standard values for the density and pressure of the earths atmosphere as a function of the distance above sea level.

ICAO Standard Atmosphere for the Earth

Height Above Sea Level (meters)	Pressure (pascals)	Density (kg/m³)
0	1.0133E+05	1.2250E+00
11,019	2.2632E+04	3.6391E-01
20,063	5.4748E+03	8.8034E-02
25,099	2.4886E+03	4.0016E-02
32,162	8.6776E+02	1.2721E-02
47,350	1.2044E+02	1.4845E-03
53,446	5.8320E+01	7.1881E-04
75,895	2.4520E+00	4.3390E-05

ICAO Standard Atmosphere for the Earth   (continued)

Height Above Sea Level (meters)	Pressure (pascals)	Density (kg/m³)
91,293	1.8150E–01	3.2130E–06
128,548	1.4510E–03	1.5660E–08
179,954	6.1900E–05	2.6550E–10
314,862	1.4470E–06	3.2790E–12

IV. Related Science Concepts

An important facet of scientific reasoning is the use of mental models to try to describe natural phenomena. For these activities use three different models to describe the earth's atmosphere.

1. Assume that the earth is surrounded by a layer of air of constant density and temperature. How thick would this atmosphere be to match the atmosphere given above? (*Hint:* How many different definitions can you use for the word "match" in the previous sentence?)

2. Assume that the earth is surrounded by an atmosphere of constant temperature and the gas molecules are attracted by a constant gravitational force. Then you can derive an expression of the pressure of the atmosphere as a function of the height, h, above sea level using the principles of hydrostatics.

EQUATIONS

$$\text{Pressure at height } h = p_0 \exp\left(\frac{-Mgh}{RT}\right),$$

where p_0 is the pressure at sea level, M is the molecular mass of air (29×10^{-3} kg/mole), g is the acceleration of gravity, R is the universal gas constant, 8.314 J/mole • K, and T is the absolute temperature in Kelvins.

3. Assume that the earth is surrounded by an adiabatic atmosphere where the pressure is given by the density, ρ raised to the γ power, where γ for the air has a value of 1.40,

$$p = k\rho^{\gamma},$$

and k is the proportionality constant. Then the hydrostatic equation yields a solution of the following form:

$$\frac{k\gamma}{\gamma(\gamma-1)}(\rho_0 - \rho) = gh,$$

From the adiabatic equation for an ideal gas, an expression for the temperature of the atmosphere as a function of height can also be derived; the following equation is obtained:

$$\frac{(T_0 - T)}{h} = \text{constant},$$

where T_0 is the temperature at sea level ($h = 0$).

Investigation 35

HEAT ENGINES

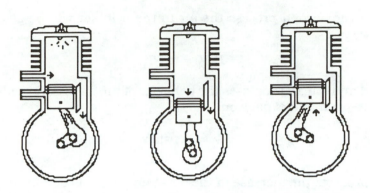

I. Wondering About...

...the performance of an internal combustion engine.

II. Investigation to Undertake

Set up your spreadsheet to use the laws of thermodynamics to compute the performance of an internal combustion engine. Use the volume of the engine cylinder and the amount of heat input from combustion as adjustable parameters. Calculate and graph the efficiency of the engine as a function of the compression ratio and the amount of energy obtained from fuel combustion. Use numerical methods to compute the work done during the portions of the cycle when the volume is changing. Also have the graphics program draw a P-V diagram for this engine.

III. Background

Use the air standard Otto cycle as an approximation for the behavior of a real engine. Assume ideal conditions: (1) the working substance, air, behaves as an ideal gas with a constant heat capacity; (2) all processes are quasi-static; and (3) there is no friction. Then the Otto cycle is composed of six simple ideal gas processes: **First**-there is quasi-static, isobaric intake of air, from zero volume to V_1 and the number of moles increases from zero to n_1 at a pressure of P_o and a temperature of T_1. **Second**-quasi-static, adiabatic compression of the air to a volume V_2 at T_2. **Third**-quasi-static, isochoric increase in temperature to T_3 and pressure brought about by the effect of burning the fuel with an input of heat Q_h. **Fourth**-quasi-static, adiabatic expansion involving a drop in temperature from T_3 to T_4. **Fifth**-quasi-static, isochoric drop in temperature to T_1 and pressure to P_o by a rejection of heat Q_c and the opening of the exhaust valve. **Sixth**-quasi-static, isobaric exhaust at atmospheric pressure while the number of moles and the volume of air drop to zero.

For cylinder volume use 0.50 liters. Try a compression ratio of 9:1. The heat of combustion for gasoline is 4.86×10^4 kJ/kg °C. Gasoline is 85% carbon and 15% hydrogen by weight.

IV. Related Science Concepts

Analyze the behavior of this engine using the Carnot cycle as a model. Note that the first and sixth processes are the exact opposites and will cancel out their contributions to engine performance and need not be considered further.

EQUATIONS

At the beginning of the second process the air in the cylinder will be given by the following equation of state:

$$P_o V_1 = n_1 R T_1.$$

For the second and fourth processes the behavior of the air is governed by the equation for an adiabatic compression and expansion:

$$PV^\gamma = \text{constant},$$

where $\gamma = 1.40$.

For the third process, the increase in the air temperature is determined from the amount of heat energy put into the system by combustion:

$$Q_h = C_v (T_3 - T_2).$$

For the fifth process the decrease in air temperature is determined from the amount of heat loss required to return the system to its original volume and temperature:

$$Q_c = C_v (T_4 - T_1).$$

The ratio of V_2 to V_1 is called the compression ratio.

ELECTRIC POTENTIAL NEAR ELECTRONS

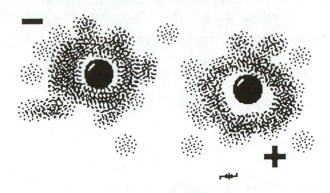

I. Wondering About...

...the electric potential near two electric charges.

II. Investigation to Undertake

Assume that your spreadsheet is an area 20 angstroms by 20 angstroms (1 angstrom = 10^{-10} m). Each cell represents one square angstrom. Use your spreadsheet to make a map of the electric potential in a plane that contains two electric charges.

Set up the spreadsheet to calculate the electric potential at the center of each cell because of the presence of two point charges centered in the cell where they are located.

III. Background

Consider the last six nonzero digits of your social security number as a way to represent the amount and location of two electric charges on your spreadsheet. Take the first two of these digits as the amount of charge in units of electronic charge (1 e = 1.6 X 10^{-19} C), positive if the digit is an even number and negative if it is an odd number. Use the last four digits to determine the (x, y) locations of the charges in the spreadsheet in columns and rows. For example, the SSN xxx-12-3456 would have $-1e$ at the location of column 3, row 4 and a $+2e$ at column 15, row 14; (i.e., (2D, 2D) – (5, 6).

IV. Related Science Concepts

The electric potential in the vacinity of a number of electric charges is just the superposition of the potential due to the charges individually. The electric potential is given by the equation:

EQUATION

$$\text{Electric potential} = \frac{Q}{4\pi\varepsilon_0 r},$$

where Q is the electric charge and r is the distance. *Hint:* Make use of the functions of your spreadsheet that will automatically read into a formula the location of a cell in terms of its row and column address.

PHYSICS TASKS

1. Draw three equipotential curves on your spreadsheet using a blue pen. Choose a low voltage value, a medium voltage value, and a high voltage value, respectively.

2. Find the regions of maximum and minimum electric field in your grid.

Show the electric field in each region as a red vector and label its magnitude (be sure to include its units).

Investigation 37

THE INCANDESCENT LIGHT BULB

I. Wondering About...

...the temperature of the filament of a 100 watt light bulb.

II. Investigation to Undertake

Consider a 100-watt light bulb designed to operate at 120 volts. Set up a spreadsheet to calculate resistance and temperature of the filament of the light bulb as a function of time after a voltage is applied to the bulb. Consider six different applied voltages: 20 V, 40 V, 60 V, 80 V, 100 V, and 120 V. Each spreadsheet should include at least five columns: time, power in, power out, temperature, and resistance. You may show other properties of the bulb in other columns as you deem appropriate. Use graphics to plot:

1. The resistance versus time for each voltage.

2. The current versus the applied voltage for the equilibrium conditions.

3. The time to reach equilibrium versus the applied voltage.

III. Background

At room temperature (about 20°C) a light bulb consists of a tungsten filament 56.4 cm long (when it is uncoiled) and 0.0063 cm in diameter. Tungsten has a specific heat of 0.128 kJ/kg °C and a density of 19,300 kg/m³. The emissivity of tungsten is 0.46. The temperature coefficient of resistivity for tungsten is about 0.0045/°C.

IV. Related Science Concepts

The input energy is the electrical energy put into the filament when the voltage is applied to the light bulb.

The increase in the internal energy of the filament is the thermal capacity of the filament.

The energy output is radiation emitted by the filament according to the Stefan-Boltzmann equation.

When the filament is heated, its resistance will change.

EQUATIONS

$$\text{Electrical power input} = \frac{V^2}{R},$$

where V is the voltage and R is the resistance. The electrical energy input is

$$\Delta E = C_v \, M \Delta T,$$

where C_v is the heat capacity of the material, m is the mass of the sample and ΔT is the temperature change of the sample.

$$R = \frac{\rho l}{A},$$

where ρ is the resistivity, l is the length of the filament, A is cross-sectional area of the filament. The resistivity, ρ, varies with temperature according to a linear relationship,

$$\rho_t = \rho_{20}(1 + \alpha \Delta T),$$

where ρ_t is the resistivity at operating temperature, ρ_{20} is the resistivity at 20°C, α is the temperature coefficient of resistivity, and ΔT is the difference between the operating temperature and 20°C.

$$\text{Emitted energy} = \varepsilon \sigma T^4,$$

where T is the absolute temperature, ε is the emissivity, and σ is the Stefan-Boltzmann constant.

The peak wavelength of emitted light

$$\lambda_m (\text{in nm}) = \frac{2.8978 \times 10^6}{T}.$$

Planck's radiation equation for blackbody emission is

$$E_\lambda = \frac{8\pi hc}{\left\{ \lambda^5 \left[\exp\left(\frac{hc}{\lambda kT}\right) - 1 \right] \right\}},$$

where c is the speed of light, h is Planck's constant, k is Boltzmann's constant, λ is the wavelength of the radiation, T is the absolute temperature, and E_λ is the energy intensity (watts/m²) emitted per unit wavelength (m).

V. Extensions

1. Use the "What...if..." capabilities of a spreadsheet to investigate:

    (a) How the diameter of the filament influences the behavior of a light bulb.
    (b) How the length of the filament influences the behavior of a light bulb.

(c) How the composition of the filament changes the behavior of a light bulb; that is, suppose the filament is made out of some other metal, such as annealed copper, or aluminum.

2. Several temperature effects may have been neglected above. What about the temperature dependence of the specific heat of tungsten or the temperature coefficient of linear expansion? How do they influence the behavior of a light bulb?

Investigation 38

ELECTRIC CIRCUITS

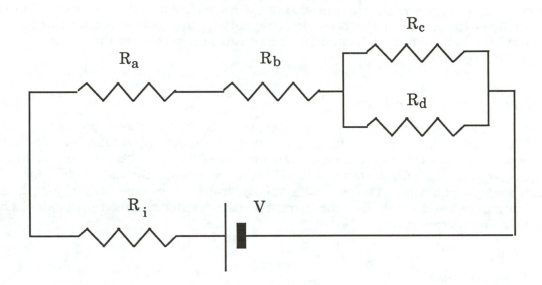

I. Wondering About...

...how the currents and voltage drops vary as resistances in the circuit are changed.

II. Investigation to Undertake

Construct a spreadsheet that calculates and displays the current through, and voltage drop across, the circuit elements in the circuit above. The resistances are to be variables that you can change by entering new numbers in cells that you designate. The voltage of the battery should also be a variable that you can select by changing the value in a cell on the sheet. The resistor R_i is to be considered as the internal resistance of the battery.

In addition to displaying the currents and voltage drops for each element, have your spreadsheet display such things as:

1. The sum of the currents in the parallel resistors.

2. The percentage of the current shared by the two resistors in parallel.

3. The sum of the voltage drops across the resistors.

4. The percentage of the total voltage drop across each resistor.

As you vary the resistances and the voltage of the battery, do the currents and voltage drops across the circuit elements behave as you expected?

You can simulate a well-used battery by allowing R_i to get large. How does the circuit respond as the internal resistance of the battery gets larger? How small does the internal resistance of the battery have to be to appear negligible? Does this answer depend on the other resistances in the circuit? How large does the internal resistance of the battery have to be to cause a drop in voltage of 20%? Does this answer depend on the other resistances in the circuit?

III. Background

Direct current circuits, such as this one, do not always behave in the way one would predict. It is very tedious to test all of the possible combinations in the laboratory in order to fully understand the behavior of such circuits. The spreadsheet version of the circuit can be very helpful in this respect.

IV. Related Science Concepts

Kirchhoff's loop and junction rules can be used to solve for the currents and voltage drops in the circuit. First the current through the battery can be calculated by computing the equivalent resistance in the entire circuit and then using Ohm's law. This current is the same as the current through R_i, R_a, and R_b. With the current through these elements, the voltage drop for each can be calculated via Ohm's law. The remaining voltage drop is across the parallel pair, R_c and R_d. Ohm's law can be used to calculate the current in each resistor and the result can be checked via Kirchhoff's junction rule.

EQUATIONS

Ohm's law states that

$$V = IR,$$

where V is the voltage drop across the circuit element, I is the current through the circuit element, and R is the resistance of the circuit element.

Kirchhoff's loop rule says that the sum of the voltage drops around a closed loop in a circuit is zero. Your textbook gives examples of its use.

Kirchhoff's junction rule says that the sum of the currents into a junction is zero. Again, your textbook gives examples of its use.

HIGH ENERGY PARTICLE STORAGE RINGS

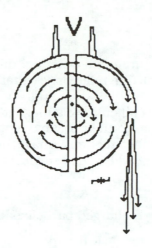

I. Wondering About...

...what kind of accelerating voltages and magnetic fields are involved in something like the Superconducting Super Collider (SSC).

II. Investigation to Undertake

Build a spreadsheet that models a large, high energy particle storage ring operation similar to that of the storage ring mode of the proposed SSC. The particles are brought up to speed by falling through a large potential difference. Magnetic fields are used to keep the particles in a circular path. Use as inputs the accelerating voltage, the particle mass and charge, luminosity (number of particles output per second, which is related to what could be called beam current), the radius of the ring, and the diameter of the beam tube. For outputs calculate: particle velocity, magnetic field required, energy in the magnetic field in the ring, and power to maintain luminosity (beam current).

III. Background

A device like the SSC has two modes of operation. In one mode, an accelerator mode, it accelerates particles up to the desired speeds. In the second mode, a storage ring mode, it maintains a number of particles at this high speed so that experimenters can divert some of the particles into collisions for study. The super collider actually does this for two kinds of particles at the same time running in opposite directions so that the collisions are even more energetic. The particles to be used are protons and antiprotons.

The SSC will actually be a circle that has been cut in half and pulled apart, and then reconnected with 4-mile straight-line segments. The 8-mile radius of the curved part is what is important to this calculation. Since the particles will naturally travel in straight lines, magnetic fields are not needed for the straight segments. The nearly 60-mile-long evacuated tube that carries the protons is 5 to 10 cm in diameter.

One can easily imagine accelerators in the straight segments that boost the particles up to 20 teraelectron volts in energy, which is the equivalent of causing these elementary charges to fall through a potential difference of 20 teravolts. If the beam had a current of 1 milliamp, then the number of particles per second would be about 6×10^{15}.

IV. Related Science Concepts

Causing a charged particle to fall through a potential difference gives it kinetic energy. The energies involved in the SSC are such that relativistic equations relate the kinetic energies to the velocities.

Charged particles moving perpendicular to a magnetic field experience a force, called the Lorentz force, which is perpendicular to their original motion and to the magnetic field. Since this force is always perpendicular to their motion we have the perfect conditions to cause uniform circular motion. The Lorentz force provides the centripetal force.

EQUATIONS

The kinetic energy acquired by a charged particle falling through a potential difference is

$$K.E. = qV,$$

where q is the charge on the particle and V is the potential difference.

The kinetic energy of a particle at relativistic speeds (near the speed of light) is

$$K.E. = \left(\frac{m_0}{\sqrt{1 - \frac{v^2}{c^2}}} - m_0 \right) c^2,$$

where m_0 is the rest mass of the particle, v is its velocity, and c is the speed of light.

A charged particle moving in a magnetic field experiences a Lorentz force, which can be expressed as

$$\mathbf{F} = q \, \mathbf{v} \times \mathbf{B},$$

where $\mathbf{F}$ is the vector force, q is the charge on the particle, $\mathbf{v}$ is the velocity, and $\mathbf{B}$ is the magnetic field vector. If the particle's velocity is perpendicular to the field, then the force will be mutually perpendicular and the expression can be used in its scalar form, $F = qvB$.

Particles moving uniformly in a circle do so under the influence of a net force called the centripetal force,

$$\sum F = m \frac{v^2}{r},$$

where m is the particle mass, v is its tangential velocity, and r is the radius of the circular path. These two expressions can be used together to develop a relationship between the variables q, v, B, m, and r.

The energy per unit volume stored in a uniform magnetic field is

$$e = \frac{B^2}{2\mu_0},$$

where μ_0 is the permeability of free space.

MAGNETIC FIELDS INSIDE A SQUARE COIL

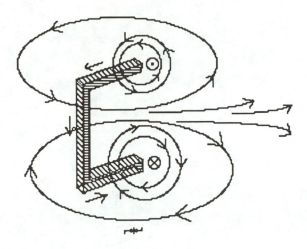

I. Wondering About...

...the uniformity of a magnetic field inside a square coil.

II. Investigation to Undertake

Set up your spreadsheet as if it were a square coil, 21 cells on a side. Put an "x" in each cell that represents part of the coil. Then you will use the Biot–Savart relationship to compute the value of the magnetic field at each cell location within the coil. Assume that each cell represents 1 cm^2 of space. Assume that the current in the coil is traveling in a counterclockwise direction, so the magnetic field is coming out of the page toward you.

III. Background

Assume that the current in the coil is 1 amp, the coil consists of only one turn of wire, and it is suspended in air. Each cell of the spreadsheet, in which you have placed an "x," represents a short piece of wire 1 cm long.

IV. Related Science Concepts

The magnitude and direction of the magnetic field in the vicinity of a current-carrying wire is given by the Biot-Savart relationship. This relationship is a differential relationship; thus the total magnetic field can be computed by adding up all the small increments of the magnetic field.

EQUATION

$$\text{Magnetic field,} \quad \Delta \mathbf{B} = \left(\frac{\mu_o}{4\pi} \right) \frac{i\,\Delta \mathbf{s} \times \mathbf{r}}{r^2}$$

STARTUP CURRENT IN AN ELECTRIC MOTOR

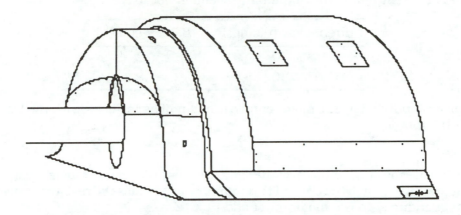

I. Wondering About...

...how the current changes as an electric motor is started.

II. Investigation to Undertake

Build a spreadsheet to calculate the current drawn by an electric motor from the time it just starts until it is up to full speed. Plot a graph of the current drawn, back emf, and battery emf versus the rotational speed up to 4000 rpm.

III. Background

When a motor is just starting, it looks to the rest of the circuit as just a long piece of wire. The current drawn is governed by the resistance of the wire and the voltage applied via Ohm's law. As it gains speed, a back emf is produced by virtue of the fact that in the motor we have the equivalent of a coil rotating in a magnetic field. The current drawn in the motor is much lower when it is at higher speeds because the back emf is proportional to rotational speed of the motor.

When a load is put on a motor, its speed reduces causing a reduction in the back emf, allowing a larger current to flow bringing more power into the motor. If the load is large enough to stall the motor, then the original startup current will flow through the motor. Such large current for a period of time is often damaging to the motor. This is why the windshield wiper motor on your car "dies," if something happens to stall it.

IV. Related Science Concepts

Let's imagine that the motor is a starter motor in a car engine. We measure the current drawn by the starter:

RPM	Current Drawn (amps)
0	250
2400	50

With this much current being drawn the battery emf is not fixed:

Current Drawn (amps)	Battery emf (V)
0	12
250	7.8

The back emf induced by the motor can be thought of as in series with the battery and the resistance of the motor.

EQUATIONS

Kirchhoff's loop rule says that the sum of the voltage drops around a circuit is zero, so the voltage drop through the battery plus the current through the motor times the resistance of the windings plus the voltage induced in the motor must add up to zero.

The emf induced in a loop rotating in a magnetic field is

$$\varepsilon = BA\omega\sin\omega t,$$

where B is the magnetic field, A is the area of the loop, and ω is the rotational velocity of the loop. Because of this the average emf is proportional to the rotational velocity.

V. Mathematical Methods

The graph of the average induced back emf versus rpm is a straight line passing through the origin. Use the spreadsheet to calculate the slope of the back emf versus rpm line.

The graph of the battery emf versus the current drawn is a straight line. Use the spreadsheet to calculate the slope of that line.

At any given rotational velocity the actual current drawn depends on the battery emf, which in turn depends on the current drawn:

$$I = \frac{\varepsilon_{battery}(I) - \varepsilon_{back}}{R}.$$

This turn of events is difficult to calculate unless one uses the iteration capabilities of the spreadsheet. For each rotational velocity,

1. Have a cell to calculate the back emf by multiplying the slope of the back emf versus rpm line times the rpm.

2. Have a cell to calculate the battery emf by multiplying the slope of the battery emf versus current drawn line by the current drawn (referring to the next cell) and adding the intercept.

3. Have a cell to calculate the current drawn using the equation just above and referring to the previous cell.

If set to iterate, the spreadsheet will solve for No. 1, then No. 2, then No. 3, and then repeat the calculations for No. 2 and No. 3 repeatedly until the limits you set on your spreadsheet are reached. Usually, the limit is a number of iterations or until the results do not change by more than a certain amount. In this case setting the amount to 0.001 will suffice. Try changing your iteration limits to see what results.

CAUSING SPARK PLUGS TO FIRE

I. Wondering About...

...how such high voltages are generated to fire the spark plugs when the electrical power sources in a car are only at 12 – 14 V.

II. Investigation to Undertake

Simulate the electrical behavior of a Kettering ignition system, which is found in many car engines. Illustrate the build-up of current in the primary circuit and magnetic field in the primary windings of the coil from the time the "points" close in milliseconds. Calculate the time for the field to decay when the points open. Using this decay of field in the primary windings of the coil, show the induced emf in the secondary windings of the coil. Use as inputs: the voltage supplied by the car's alternator, the capacitance, inductance, and resistance in the primary circuit, the number of turns and length of the primary windings of the coil, and the number of turns and diameter of the secondary windings of the coil. Adjust your inputs to see if you can get a "hot" ignition, one capable of 25 kV or more.

Normally, the primary has enough time to build its field all the way up before the points open again. At high rpm's this does not happen; have your spreadsheet calculate the induced emf if the points open 1 msec after the points close, repeat for 2 msec, and so on. If at least 8 kV are required to fire the plug, how long   must the points wait before opening? This poses an upper limit on the rpms for reliable operation.

III. Background

In the normal spark ignition engine a large voltage is induced across a gap between electrodes in a spark plug to produce a spark that delivers 0.05 to 0.10 J to the air–fuel mixture in the cylinder. The resulting explosion releases energy to enable the engine to do work. The large voltage is induced by causing the magnetic field in one coil, called the primary windings, to drop rapidly. This sudden change in magnetic field induces a large emf in another coil, called the secondary windings, which is completely surrounded by the primary windings. The sudden drop in magnetic field of the primary windings is accomplished by a rapid cessation of current in those windings. This combination of coils, somewhat like a transformer, is called an induction coil.

Normally, voltages in excess of 18,000 V are available to the spark plug. How can this be possible every time a plug fires as many as 12,000 times per minute when the alternator only produces about 14 V when the engine is running?

IV. Related Science Concepts

The current in the primary windings is controlled by a switch, called the points, and a capacitor. When the points are closed, the primary current builds up at a rate determined by the inductance of the primary windings and the resistance in the primary portion of the circuit. With this build-up of current is an accompanying build-up of magnetic field in the primary windings. We will be ignoring the mutual induction of the coils in this investigation. (You can choose not to ignore it if you wish.) When the points open, the current suddenly begins to drop; this drop in current is controlled by a capacitor, which is across the open switch and the inductance of the primary windings. The capacitor plus coil form an oscillating circuit in which the energy moves back and forth between the capacitor and the coil. It takes one quarter of the natural period of this circuit for the current to fall to zero from its value when the points opened. In this same period of time the field from the primary windings drops to zero. This results in a flux change in the secondary windings, which due to its high rate and the large number of turns gives rise to a large emf.

Changes in the properties of the circuit can result in greater induced emf, allowing the plugs to fire more reliably through a variety of conditions. Unfortunately, most of the changes are either expensive (upping the turns ratio) or result in reduced life of the components (increasing the primary current or reducing the primary capacitance). One approach that is being used on an increasing number of cars is an electronic ignition. This circuit uses the points or some other less mechanical device to trigger when to act. The circuit gets the current in the primary up quicker and pulls it down faster than the standard primary. The result is lower maintenance and higher available voltage to the plugs.

EQUATIONS

The current in the primary when it reaches steady state is determined by the supplied voltage and the resistance in the primary according to Ohm's law:

$$V = IR.$$

The current builds up to this value in an inductance dominated circuit according to the expression

$$I = I_0\left[1 - \exp\left(\frac{-Rt}{L}\right)\right],$$

where I_0 is the steady state current, R is the resistance in the circuit, and L is the inductance in the circuit.

The field in a solenoid is

$$B = \frac{\mu_0 NI}{l},$$

where μ_0 is the permittivity of free space, N is the number of turns in the solenoid, I is the current in the solenoid, and l is the length of the solenoid.

The natural period of an *LC*-dominated circuit is

$$T = 2\pi \sqrt{LC} .$$

When a field changes in a solenoid, the magnitude of the induced emf is

$$\text{emf} = NA\frac{\Delta B}{\Delta t},$$

where *N* is the number of turns in the coil, *A* is the area of a coil, ΔB is the change in the magnetic field, and Δt is the time for the change to occur.

Here is a reasonable set of parameters to start with:

Supply voltage (V)	14
Turns in primary, N_p	180
Turns ratio, N_s/N_p	100
Resistance in primary, R_p (Ω)	2 – 4
Inductance in primary, L_p (mH)	5 – 12
Capacitance in primary, C_p (μF)	0.20
Length of primary solenoid, l_p (cm)	5
Radius of secondary coil, r_s (cm)	2

For additional investigations use these parameters:

Inductance in secondary, L_s (H)	70 – 100
Capacitance in secondary, C_s (pF)	60 – 80

Investigation 43

TUNING YOUR RADIO

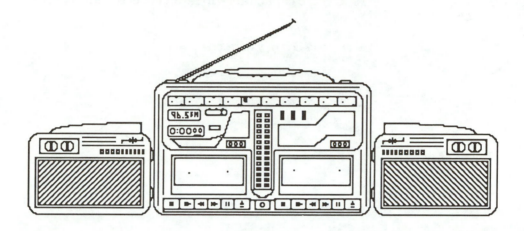

I. Wondering About...

...how turning the tuning knob on a radio changes the station you hear.

II. Investigation to Undertake

Treat the tuning circuit of your radio as a simple *RLC* circuit and examine its frequency response as you change the capacitance in the circuit. A radio that has a very precise tuning circuit will have a very narrow response peak as the capacitance is changed. Set up a spreadsheet to compute the impedance, capacitive reactance, inductive reactance, and power for various frequencies and various values of capacitance. Plot your results as graphs of power versus capacitance and power versus frequency.

III. Background

A typical AM radio station will have a broadcast frequency of about 1000 kilohertz, or 6.28×10^6 radians per second. Select typical values for resistance (about 10 ohms), inductance (about 2 mH) and capacitance (about 50 pF) to give your *RLC* circuit a resonance frequency of 1000 kH. Assume an incident power of one watt.

IV. Related Science Concepts

For a series *RLC* ac circuit, the impedance of the circuit varies as the frequency is changed. The behavior of the ac circuit is given by the following equations:

EQUATIONS

$$\text{Voltage} = \text{current X impedance or } V = IZ$$

99

$$Z = \sqrt{R^2 + (X_c - X_L)^2}$$

$$X_c = \frac{1}{\omega C} \text{ and } X_L = \omega L,$$

where ω is the frequency in rad/s, C is the capacitance in farads, L is the inductance in henries, and R is the resistance in ohms.

$$\text{Power} = VI \cos \phi, \text{ where } \cos \phi = \left(\frac{R}{Z}\right).$$

DOPPLER RADAR SPEED MEASUREMENTS

I. Wondering About...

...how quickly Doppler radar can detect a vehicle's speed.

II. Investigation to Undertake

Use your knowledge of kinematics, Doppler shift, and beats to determine how long it might take such a radar speed detection unit to determine your speed and how far you would travel during that time. Set up your spreadsheet so that various speeds can be entered to see the results at these speeds. Establish columns calculating relative velocity, frequency emitted, frequency received, beat frequency, time for three beats, and distance moved during three beats. Does the speed that you travel with respect to the radar unit have any effect on the time for the device to determine your speed or the distance that you travel while it is determining your speed?

III. Background

Most of us have seen police using radar speed detectors. This equipment has been on our roads and highways for several decades. One wonders whether it is possible to slow down quickly enough when one discovers that such equipment is being used. Is it realistic to imagine that such a maneuver is possible or should one see to it that speed limits are obeyed?

IV. Related Science Concepts

Doppler radar units send out electromagnetic waves in the 10-cm wavelength range, approximately. (You may want this to be another variable in your spreadsheet.) These waves reflect off of objects and return to the radar unit. The incoming waves and a sample of the outgoing waves are added. If the returning waves happen to be frequency shifted, then beats will occur between the waves from the two different sources. Indeed any waves reflected from objects moving with respect to the radar unit will be Doppler shifted. In fact, by the time the waves are detected by the radar unit they will be "shifted" twice, once at reflection and again on detection by the radar unit. A reflecting object "hears" the original waves shifted because it is moving with respect to the radar unit. This object reflects the waves as it "hears" them. The radar unit "hears" them shifted again because it is moving with respect to the reflecting object.

101

These frequency-shifted waves, when added to a sample of the original waves, cause beats. The frequency of the beats is equal to the difference between the frequencies. If the original frequency is known, then the incoming frequency can be calculated. The incoming frequency is higher if it is due to reflections from an oncoming vehicle, and lower if it is due to reflections from a vehicle moving away.

EQUATIONS

The beat frequency is equal to the difference between the two frequency waves that produce the beats:

$$f_{beat} = |f_1 - f_2|.$$

When the source of a wave is moving with respect to an observer of the wave, the frequency that the observer "sees" is

$$f_0 = f\frac{(v + v_0)}{(v - v_s)},$$

where f is the frequency that the source emits, v is the speed of the wave in the medium in which it is traveling, v_0 is the speed of the observer with respect to a stationary point along the side of the road, and v_s is the speed of the source with respect to a point stationary along the side of the road. The values of v_s and v_0 are taken to be positive if the source and observer are approaching each other, and negative if they are moving away from each other.

THE BRIGHTNESS OF LIGHT ON MY DESK

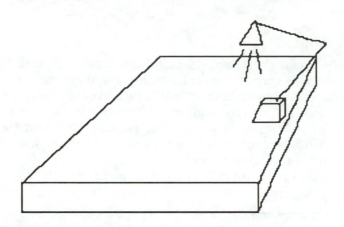

I. Wondering About...

...how much brighter is the light directly under the study lamp compared to the brightness at the farthest corner of my desk.

II. Investigation to Undertake

Set up a spreadsheet as a grid map of the top of your desk and compute the relative brightness of the light shining from a small study lamp onto your desk top. The number that appears in each cell of the spreadsheet will represent the brightness of the light at that location in, say, 1 in. by 1 in. squares.

III. Background

Let's imagine that your desk top is 30 in. wide and 4 ft long. The light is a small study lamp with a bright point source filament and clear glass so that the position of the filament can be measured. Assume that the filament is 18 in. above the table and directly over a point 3.5 in. from a long edge of the table and 24.5" from the left edge (The filament is directly over the 4th cell down and the 25th cell from the left edge.) of the table.

IV. Related Science Concepts

The brightness of light can be defined as the light energy falling on a unit area of a surface. Then, the brightness of light from a point source varies inversely as the square of the distance from the source to the illuminated surface area. If the brightness of the light directly under the lamp is some value, B_o, then the brightness at other locations can be given in terms of a fraction of that intensity. The brightness, B, at any other location can be written as

$$B = B_0 \left(\frac{d_o^2}{d^2} \right),$$

where d_o is the distance from the lamp to the point directly beneath it and d is the distance from the lamp to the center of a particular location, or area, of interest. This equation can be used to calculate the relative brightness of light on a surface as the distance to the surface changes.

V. Mathematical Methods

The rectangular gird of a spreadsheet can be thought of as a map of a surface, in this investigation as the top of your desk. Each spreadsheet cell can represent an area on the desk top, say 1 in. by 1 in.. Then the center of each spreadsheet cell is 1 in. in each direction from the center of its neighbors. You can think of the row and column numbers as the inch distances in two perpendicular directions on the top of your desk. Your spreadsheet has functions that will automatically put the location of any cell of the spreadsheet into a formula for you. These are often known as row() and column(), respectively.

The distance from the lamp to any cell (that is, square area on the desk top) can be computed using the Pythagorean theorem:

$$\text{Distance} = \sqrt{\left| 18^2 + (\text{row}(\ ) - \text{lamp row})^2 + (\text{column}(\ ) - \text{lamp column})^2 \right|},$$

where 18 in. is the height of the lamp above the desk, *lamp row* is the number of the row of the cell in the spreadsheet that represents an area directly under the lamp, and *lamp column* is the number of the column of the cell in the spreadsheet that represents an area directly under the lamp. All distance calculations may be done using inches.

This equation can be used in the spreadsheet to calculate the distance from the lamp to the area on the desk top that is represented by that cell. Then the number in each cell of the spreadsheet can represent the relative brightness of the light at that location on the desk top.

HUMAN VISION

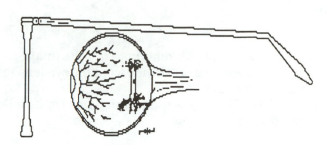

I. Wondering About...

...the kind of eyeglasses needed to correct various vision defects.

II. Investigation to Undertake

Develop a spreadsheet into which you can enter the near-point and far-point distances for a person's clear vision and compute the kind and power of eyeglass lenses the person will need to have normal vision.

III. Background

The human eye is an optical system of unsurpassed versatility. The light-focusing system of the eye is able, for normal visual, to bring objects from infinity as well as from 20 cm into focus. Most of the focusing power of the eye is accounted for by the ciliary muscles that control the curvature of the cornea.

    The distance from your eye to the nearest object you can see clearly is called your near-point distance. The normal near-point distance is considered to be 25 cm.

    The distance from your eye to the farthest object you can see clearly is called your far-point distance. The normal far point is taken as infinity.

    If your near-point is farther than 25 cm from your eye, or if your far-point is closer to you than infinity, then your vision is defective and needs to be corrected by the use of eyeglasses.

IV. Related Science Concepts

The power of a lens is a measure of its ability to bend rays of light. A lens of high power has a short focal length. The numerical value of the power of a lens is measured in diopters and is

105

calculated by taking the reciprocal of the focal length of the lens in meters. For example, a converging lens with a 5-cm focal length is called a lens of +20 diopters. A minus sign represents a diverging lens.

The simple lens formula and a knowledge of the properties of the images formed by converging and diverging lenses can be used to compute the lenses required to correct various kinds of defective vision.

EQUATIONS

$$\frac{1}{p} + \frac{1}{q} = \frac{1}{f},$$

where p is the object distance, q is the image distance, and f is the focal length of the lens. For the normal eye, the distance from the eye lens to the retina on which the image must be focused is about 3 cm.

Use this equation and the data about the eye to set up your spreadsheet to compute the proper eyeglass lenses.

SIMPLE LENSES

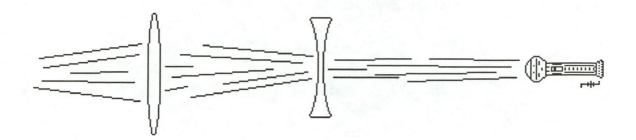

I. Wondering About...

...images as seen through glass lenses.

II. Investigation to Undertake

Use the thin lens equations to set up a spreadsheet that will compute the location and size of an image for the various distances an object is from the lens. Use a graph to show the size of the image as a function of distance from the lens. Do this exercise for both converging and diverging thin lenses.

III. Background

The basic element of many optical devices, such as microscopes, binoculars, and telescopes, is a thin lens. Even complex systems that consist of a variety of specialized, thick lenses can usually be understood by using the principle of superposition and the properties of thin lenses.

IV. Related Science Concepts

The simple, thin lens formula and a knowledge of the properties of the images formed by converging and diverging lenses can be used to compute the image location and magnification of a lens. The thin lens equation is as follows.

EQUATIONS

$$\frac{1}{p} + \frac{1}{q} = \frac{1}{f},$$

where p is the object distance, q is the image distance, and f is the focal length of the lens (positive for a converging lens and negative for a diverging lens).

The magnification is given by the ratio of the image size to the object size, but since the sizes of the object and image are proportional to their distances from the lens, the magnification can be written as a ratio of distances:

$$\text{Magnification} = -\frac{q}{p},$$

where a minus sign is used to indicate that the image is inverted from the object.

Use these equations and a variety of object distances to compute the various image locations and magnifications for typical thin lenses.

Have your spreadsheet program plot the image locations and the magnifications for your lenses as functions of the object distances.

DOUBLE–SLIT INTERFERENCE PATTERNS

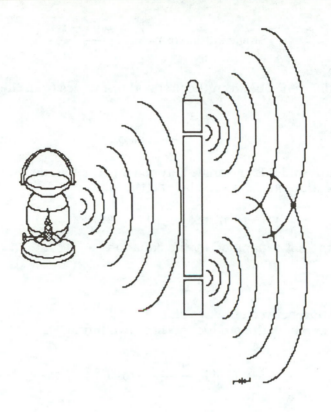

I. Wondering About...

...the double-slit interference pattern and how it changes with slit width and slit separation.

II. Investigation to Undertake

Set up your spreadsheet to calculate the light intensity as a function of distance from the center of the central bright spot of a double-slit interference pattern. In addition, have the spreadsheet calculate the difference in path lengths from the slits to each point where the intensity is calculated. Arrange your spreadsheet so that you can treat the slit width, the slit separation, and the wavelength as adjustable parameters. Use the graphics portion of your spreadsheet to plot several different interference patterns for different slit widths and slit separations.

III. Background

In 1801, Thomas Young placed the wave model for light on a firm experimental basis through his observation of the interference patterns from light that passed through two pinholes. This experiment is now regularly used as a demonstration of the wave nature of light.

IV. Related Science Concepts

The interference pattern intensity can be calculated by combining the interference effect of a single slit and diffraction effect that occurs for two slits.

EQUATIONS

$$\text{Single–slit intensity} = I_o \frac{\sin^2\left(\frac{\phi}{2}\right)}{\frac{\phi^2}{4}},$$

where ϕ is the phase difference between the first and last waves from the top and bottom of the slit,

$$\phi = \frac{2\pi}{\lambda} a \sin \varphi,$$

where a is the slit width and φ is the angle between a perpendicular line between the two slits and the line to the location on the screen behind the slits.

$$\text{Two–slit intensity} = I(\delta) = 4 I_o \cos^2\left(\frac{\delta}{2}\right)$$

where δ is related to the slit separation by the equation

$$\delta = \frac{2\pi}{\lambda} d \sin \varphi,$$

where d is the separation between the two slits.

The resulting intensity is the product of these two terms

$$I = 4 I_o \frac{\sin^2\left(\frac{\phi}{2}\right)}{\frac{\phi^2}{4}} \cos^2\left(\frac{\delta}{2}\right).$$

Do the locations of maximum destructive interference according to intensity correspond to the locations of maximum destructive interference according to path-length differences? For constructive interference?

RUTHERFORD SCATTERING

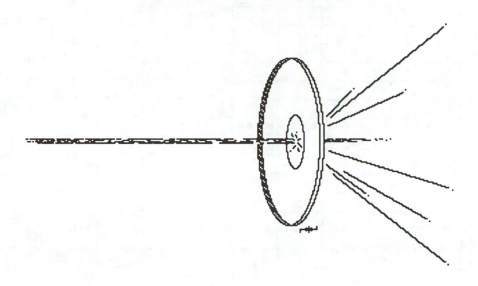

I. Wondering About...

...the scattering of alpha particles from a thin gold foil.

II. Investigation to Undertake

Set up your spreadsheet to look like a cross section of a scattering chamber, with a beam of alpha particles coming in from the left side of the spreadsheet, say along row 11, and striking a target, say in column 11 (or K), rows 10, 11, and 12 . Assume that the cells located about 10 cells from K11 are detectors that can count alpha particles. Assume that all of the detectors have the same cross-sectional area, are located in the center of the cells, and are facing perpendicular to the target. Use the results of Rutherford's work to compute the number of alpha particles per incoming 100 million particles that would be scattered into each detector (or cell). See the following example of how to set up your spreadsheet with each of the 56 detectors numbered.

III. Background

Ernest Rutherford's explanation for the alpha particle scattering experiments carried out by his students, H. W. Geiger and E. Marsden, led to the nuclear model for the atom in 1911. They scattered a beam of 5-MeV alpha particles from a gold foil about 10^{-4} cm thick. How many particles did they detect that were scattered through large angles?

Since that time, scattering experiments have been a major experimental tool by which physicists probe the secrets of nature.

	A	B	C	D	E	F	G	H	I	J	K	L	M	N	O	P	Q	R	S	T	U
1							10	11	12	13	14	15	16	17	18						
2						9										19					
3					8												20				
4				7														21			
5			6																22		
6		5																		23	
7	4																				24
8	3																				25
9	2																				26
10	1										[										27
11	56	•	•	•	•	•	•	•	•	•	[	target									28
12	55				beam						[										29
13	54																				30
14	53																				31
15	52																				32
16		51																		33	
17			50																34		
18				49														35			
19					48												36				
20						47										37					
21							46	45	44	43	42	41	40	39	38						

IV. Related Science Concepts

We can imagine an energetic alpha particle, a helium nucleus with charge, +2e, suffering a near collision with a massive gold nucleus. The Coulomb repulsion is enough to keep them from actually making contact, but the alpha particle makes a large change in its direction of motion. Such a trajectory would not be hard to calculate using what you have learned in physics so far. Unfortunately, in an actual experiment a large number of alpha particles have to pass through a wall of gold foil many atomic layers thick. The derivation of the number of alpha particles that come out at various angles is usually left to upper division texts. The result of such a derivation follows.

EQUATIONS

The number of particles that come out in a small angular range is determined to be

$$\frac{\Delta N}{\Delta \theta} = 2\pi N n \left[\frac{K}{2\,mv_o^2}\right]^2 \frac{\sin\theta}{\sin^4\left(\frac{\theta}{2}\right)},$$

where ΔN is the number of particles scattered into a small angular range, $\Delta\theta$ is the small angular range, N is the total number of particles scattered at all angles, mv_o is twice the initial kinetic energy of the alpha particles, and θ is the angle at which the particles are scattered from the original direction,

$$K = \frac{2Ze^2}{4\pi\varepsilon_o},$$

where Z is the atomic number of the atoms in the foil (47 for Au) and e is the elementary charge. (Note that the 2 in this expression reflects the fact that the incident particles are alphas with a charge, $+2e$.) Then

$$n = N_o \frac{\rho \pi r^2 t}{M},$$

where n is the number of scattering nuclei, N_o is Avogadro's number, ρ is the mass density of the atoms in the foil (10.5×10^3 kg/m^3 for Au), r is the radius of the incident beam of alpha particles (say 1 cm), t is the thickness of the foil, and M is the atomic mass of the atoms in the foil (107.7 for Au).

These expressions can be used to determine the numbers of particles scattered into each of the cells designated as a number detector in your spreadsheet. Make a graph of the logarithm of the number of particles versus scattering angle. What happens as you change foil-thickness? Foil material? Alpha particle energy?

RADIOACTIVITY

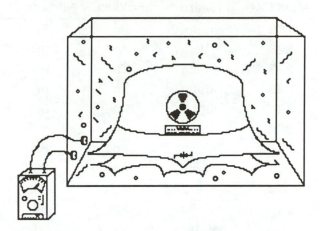

I. Wondering About...

...the populations of radioactive elements.

II. Investigation to Undertake

Set up your spreadsheet, using the equation for radioactive decay, to track the relative amounts of the various elements in a radioactive decay series. Examine what happens as time passes. Determine how this behavior can be used to estimate the age of the universe. What assumptions must you make to carry out such an estimate?

III. Background

Almost all of the naturally occurring radioactive elements lie in the range of atomic numbers from $Z = 81$ to $Z = 92$. These elements have been grouped into three decay series: the uranium series, the thorium series, and the actinium series, named for the parent element. Any one of the radioactive elements can be traced back through a series of transformations to its parent element.

IV. Related Science Concepts

Naturally occurring radioactive elements decay with a typical lifetime. The number of any species at time, t, $N(t)$, is given by an exponential decay function.

EQUATIONS

$$N(t) = N_0 \exp\left(\frac{-0.693t}{\tau}\right),$$

115

where τ is the half-life, N_o is the number at $t = 0$, and t is the time. (*Hint:* the use of a conditional statement may be required to avoid error messages in your spreadsheet.)

The decay scheme for the Thorium series follows.

Thorium Series

Element Symbol	Atomic Number (Z)	Atomic Mass (M)	Decay Means	Decay Product	Atomic Number (Z)	Atomic Mass (M)	Half-life
Th	90	232	alpha	Ra	88	228	1.39E10y
Ra	88	228	beta	Ac	89	228	6.7y
Ac	89	228	beta	Th	90	228	6.13h
Th	90	228	alpha	Ra	88	224	1.9y
Ra	88	224	alpha	Em	86	220	3.64d
Em	86	220	alpha	Po	84	216	54.5s
Po	84	216	alpha	Pb	82	212	0.16s
Pb	82	212	beta	Bi	83	212	10.6h
Bi	83	212	beta	Po	84	212	60.5m
Po	84	212	alpha	Pb	82	208	3E-7s
Pb	82	208	stable				

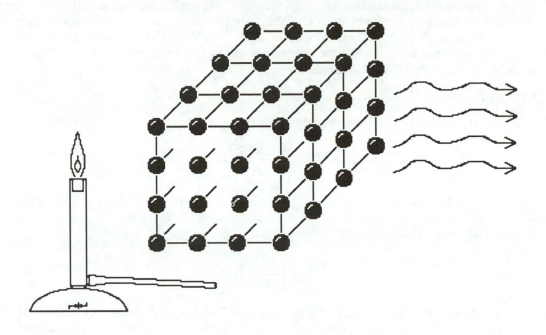

Investigation 51

THE LOW–TEMPERATURE HEAT CAPACITY OF METALS

I. Wondering About...

...how much of the low-temperature heat capacity of a metal is due to the electrons and how much is due to lattice vibrations.

II. Investigation to Undertake

Set up a spreadsheet to calculate the total heat capacity of a metal as a function of temperature at low temperatures (0 to 40 K). Compute the fraction of the heat capacity that is electronic. Graph the results for several metals, showing the total heat capacity, the electron contribution, and the lattice vibration contribution.

III. Background

The dependence of the heat capacity, or internal kinetic energy, of a metal on temperature was one of the crucial experiments in the beginning of modern solid state theoretical physics. The low-temperature (<40K) behavior of the thermal properties of metals did not agree with the simple free-electron gas model that was used to explain the room temperature and above thermal behavior of metals. This problem attracted the attention of such now famous scientists as Albert Einstein and Peter Debye. The major breakthough came with the consideration of lattice vibrations and the development of theoretical models to predict the lattice vibration contributions to the heat capacity of a metal.

117

IV. Related Science Concepts

Both electrons and vibrations of the lattice can have kinetic energy. How much of the internal kinetic energy of a metalic solid is partitioned to the lattice is determined by the lattice stiffness that is related to a parameter of the metal called the Debye temperature. How much of the internal kinetic energy of a metalic solid is partitioned to the electrons is related to a property of the metal called the Fermi energy. The contribution to the total heat capacity, C, by the lattice vibrations is proportional to the cube of the absolute temperature. The electronic contribution to the total heat capacity is directly proportional to the absolute temperature. Hence the relative amounts of the electronic and vibronic contributions to the heat capacity of a metal changes dramatically as the temperature increases.

EQUATIONS

The lattice contribution to the heat capacity of a metal is

$$C_L = \left(\frac{12\pi^4 R}{5\Theta_D^3} \right) T^3,$$

where Θ_D is the Debye temperature of the material and R is the universal gas constant (2 cal/mol-K).

The electron contribution to the heat capacity is

$$C_e = \left(\frac{\pi^2 R k_B}{2E_F} \right) T,$$

where E_F is the Fermi energy for the material and k_B is Boltzmann's constant (1.38×10^{-23} J/K).

The total heat capacity, C, at low temperatures is the sum of these two contributions.

Element	E_F (eV)	Θ_D(K)	Element	E_F (eV)	Θ_D(K)
Ag	5.5	225	In	8.6	108
Al	11.6	428	Na	3.1	158
Au	5.5	165	Mg	7.1	400
Ba	3.6	110	Pb	9.4	95
Be	14.1	1440	Sn	10.0	200
Cs	1.6	38	Zn	9.4	327
Cu	7.0	343			

Note: 1 eV = 1.6 X 10^{-19} J.

Investigation 52

THE COLOR OF A HOT FILAMENT

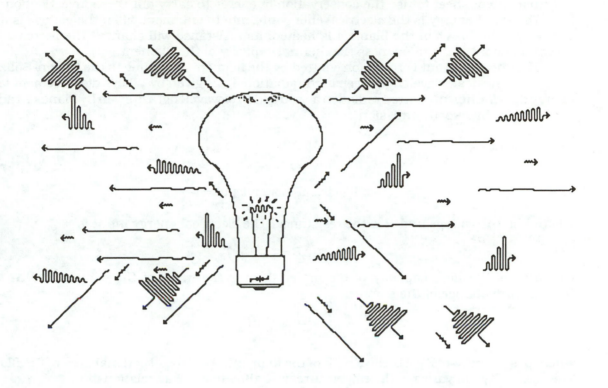

I. Wondering About...

...the color of a hot filament and how much of its energy is emitted in the visible region of the electromagnetic spectrum.

II. Investigation to Undertake

Consider a 100-watt light bulb designed to operate at 120 V. Set up a spreadsheet to calculate filament resistance, filament temperature, maximum wavelength λ_m of the brightest light emitted, and the relative amount of the total emitted energy that falls in the visible region (380 nm – 750 nm) of the electromagnetic sprectrum as a function of time after a voltage is applied to the bulb. Consider six different applied voltages: 20 V, 40 V, 60 V, 80 V, 100 V, and 120 V. Each spreadsheet should include at least seven columns: time, power in, power out, temperature, resistance, λ_m, and the relative amount of visible energy.

III. Background

At room temperature (about 20°C) a light bulb consists of a tungsten filament 56.4 cm long (when it is uncoiled) and 0.0063 cm in diameter. Tungsten has a specific heat of 0.128 kJ/kg °C and

119

a density of 19,300 kg/m³. The emissivity of tungsten is 0.46. The temperature coefficient of resistivity for tungsten is about 0.0045/°C.

IV. Related Science Concepts

Set up a spreadsheet to use the conservation of energy to carry out these investigations.

The input energy is the electrical energy put into the filament when the voltage is applied to the light bulb. When the filament is heated, its resistance will change. The increase in the internal energy of the filament is the thermal capacity of the filament.

The energy output is radiation emitted by the filament according to the Stefan-Boltzmann equation. The peak wavelength of emitted radiation, l_m, is given by Wien's displacement law. The energy distribution of emitted radiation as a function of wavelength is given by Planck's radiation equation for blackbody emission.

EQUATIONS

$$\text{Electrical power input} = \frac{V^2}{R},$$

where V is the voltage and R is the resistance. The electrical energy input is

$$\Delta E = C_v \, M \Delta T,$$

where C_v is the heat capacity of the material, m is the mass of the sample and ΔT is the temperature change of the sample.

$$R = \frac{\rho l}{A},$$

where ρ is the resistivity, l is the length of the filament, A is cross-sectional area of the filament. The resistivity, ρ, varies with temperature according to a linear relationship,

$$\rho_t = \rho_{20}(1 + \alpha \Delta T),$$

where ρ_t is the resistivity at operating temperature, ρ_{20} is the resistivity at 20°C, α is the temperature coefficient of resistivity, and ΔT is the difference between the operating temperature and 20°C.

$$\text{Emitted energy} = \varepsilon \sigma T^4,$$

where T is the absolute temperature, ε is the emissivity, and σ is the Stefan-Boltzmann constant. The peak wavelength of emitted light

$$\lambda_m(\text{in nm}) = \frac{2.8978 \times 10^6}{T}.$$

Planck's radiation equation for blackbody emission is

$$E_\lambda = \frac{8 \pi h c}{\left\{ \lambda^5 \left[\exp\left(\frac{hc}{\lambda \, kT} \right) - 1 \right] \right\}},$$

where c is the speed of light, h is Planck's constant, k is Boltzmann's constant, λ is the wavelength of the radiation, T is the absolute temperature, and E_λ is the energy intensity (watts/m²) emitted per unit wavelength (m).

RELATIVISTIC SPEEDS OF ELECTRONS
IN A LINEAR ACCELERATOR

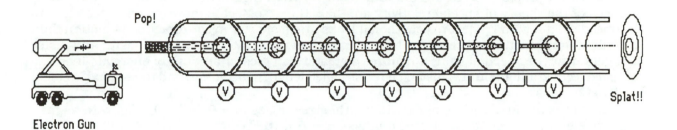

Pop!

Electron Gun

Splat!!

I. Wondering About...

...how the world looks to electrons as they go faster and faster in a linear accelerator.

II. Investigation to Undertake

Set up a spreadsheet that shows the kinetic energy of the electrons, their speed, their relativistic mass, the ratio of their relativistic mass to their rest mass, and the apparent length of the whole accelerator at each 100-m-long section of the linear accelerator. Assume that the accelerator consists of a total of 32 equivalent sections and the acceleration voltage applied across each section is the same.

At the top of the spreadsheet label a cell as the accelerating voltage applied per meter. You should set up the spreadsheet so that by varying the voltage in this cell you can see what it is like to turn up the accelerating voltage in the accelerator. Beginning with an accelerating potential of 100 volts per meter, and increasing it by a factor of two each time, examine the behavior of electrons as they go faster and faster. Finally, find the voltage, V_{max}, which makes the 3200-m accelerator look about 0.66 m long to the electrons at the exit of the accelerator. Then arrange your spreadsheet to show the results for the voltage V_{max} that you find and two graphs: one graph showing the electron speed versus distance along the accelerator for three accelerating voltages, V_{max}, $(1/2)V_{max}$, and $(0.1)V_{max}$, and the other graph showing the apparent length of the accelerator versus distance along the accelerator at the same three voltages.

III. Background

Sidney Drell, a physicist at Stanford University, in a film titled "Einstein's Universe," indicates that the electrons accelerated in the Stanford Linear Accelerator (SLAC) are going so fast at the end of a run that the 2-mile accelerator appears to be only 2 ft long to them. Translated into SI units, the accelerator is about 3200 m long and to the electrons at the end of the run it looks only about 0.66 m long.

121

IV. Related Science Concepts

The theory of special relativity formulated by Albert Einstein in 1905 suggests an explanation of this strange statement. The foundation of the theory is based on two statements of apparent fact: (1) the basic laws of physics hold for all observers and (2) the speed of light is the same for all observers. The consequences of these two statements taken together are worked out in the theory using no more sophisticated mathematics than geometry. English translations of Einstein's original paper are found in most libraries in books on relativity and can be read by anyone with a preparation in algebra and geometry from high school. Two important consequences of the theory of special relativity are that (1) if an object is moving at a speed close to the speed of light with respect to you, then it looks much shorter in the direction of motion than it would if it were sitting at rest next to you and (2) as you make an object go faster and faster as it gets near the speed of light, it requires a greater and greater amount of energy to increase its speed a constant amount. This first effect is called length contraction and the second is called relativistic mass.

Electrons are extremely small particles, their rest mass $m_0 = 9.11 \times 10^{-31}$ kg, with a small negative electric charge, their electric charge $q = -1.6 \times 10^{-19}$ C.

Electrons can be sped up by electric potentials applied along sections of the accelerator. To accelerate the electrons, the potential is applied across each of these sections as the electrons pass through it. The electrons are pushed ahead by a continually accelerating electric force, but after the electrons have passed a section the electric potential in that section is turned off and the potential in the next section is not turned on until the electrons get to it. In effect, the electrons are riding the "crest of a wave" like a surfer.

EQUATIONS

The kinetic energy of a relativistic particle is

$$K. E. = mc^2 (\gamma - 1),$$

where

$$\gamma = \frac{1}{\sqrt{\left[1 - \left(\frac{v}{c}\right)^2\right]}},$$

is a number equal to 1 for a speed of zero and is always larger than 1, for example, if an electron is traveling at one half the speed of light γ is 1.15.

The relativistic mass of a particle is equal to $m_0\gamma$, so a particle moving at one half the speed of light has a relativistic mass 1.15 times its rest mass.

The apparent contracted length of an object not in the particle's rest frame is

$$L = \frac{L_0}{\gamma},$$

where L_0 is the proper length of the object (its length when it is sitting at rest next to you.)

The kinetic energy given to an electron accelerated by an electric potential is

$$K. E. = qV,$$

where q is the charge on the particle and V is the size of the electric potential measured in volts. The speed of light is $c = 3.0 \times 10^8$ m/s.

These equations can be used in the spreadsheet to compute the speed, relativistic mass, and contracted length as a function of the accelerating electric potential.

3 6 9 1 G

3